Engineering Design

Project Guidelines

Alan D. Wilcox, Ph.D., P.E.
Bucknell University

Prentice-Hall, Inc.
Englewood Cliffs, New Jersey 07632

Library of Congress Cataloging-in-Publication Data

Wilcox, Alan D. (date)
Engineering design.
Bibliography: p.
Includes index.
1. Engineering design. 2. Industrial
project management. I. Title.
TA174.W44 1987 620'.00425 86-22565
ISBN 0-13-277989-7

Editorial/production supervision and
interior design: Linda Zuk, WordCrafters Editorial Services, Inc.
Cover design: 20/20 Services, Inc.
Manufacturing buyer: Rhett Conklin

Printed in the United States of America
10 9 8 7 6 5 4 3 2 1

ISBN 0-13-277989-7 025

Prentice-Hall International (UK) Limited, *London*
Prentice-Hall of Australia Pty. Limited, *Sydney*
Prentice-Hall Canada Inc., *Toronto*
Prentice-Hall Hispanoamericana, S.A., *Mexico*
Prentice-Hall of India Private Limited, *New Delhi*
Prentice-Hall of Japan, Inc., *Tokyo*
Prentice-Hall of Southeast Asia Pte. Ltd., *Singapore*
Editora Prentice-Hall do Brasil, Ltda., *Rio de Janeiro*

Over an entire lifetime
one might have many fine acquaintances.
However, during that same period,
one might have only several true friends.
They are true always and forever,
and this book is dedicated to my two best:
Royce F. Crocker and Rose Marie M. Crocker.

Contents

About the Author

Alan D. Wilcox is associate professor of Electrical Engineering at Bucknell University. Since coming to Bucknell in mid-1983, he has taught courses in computer programming, digital logic, and computer system design.

Professor Wilcox received his Ph.D. in electrical engineering in 1976, his M.E.E. in 1974, and his M.B.A. in 1972, all from the University of Virginia. He received his B.E.E. from Rensselaer Polytechnic Institute in 1965. He is a licensed Professional Engineer and has been involved with computers since the mid-1960s. His current technical interests are microprocessor architecture, troubleshooting, and multitasking-multiprocessing computer systems. He is a member of the IEEE, ACM, ASEE, and Eta Kappa Nu.

Before joining Bucknell, Dr. Wilcox worked in Virginia with E-Systems as a principal engineer doing advanced development of digital speech-enhancement hardware. Earlier, he was a project manager with Weston Controls and was responsible for microprocessor software development for nuclear instrumentation.

Dr. Wilcox can be contacted at 60 South 8th Street, Lewisburg, PA 17837. Telephone (717) 523-0777.

Preface

ABOUT THIS BOOK

There is a large gap between the engineering design usually studied in school and the actual practice of engineering in industry. This book is unique because it integrates the principles of engineering design with practical hands-on experience in the real world. Its purpose is to provide a unified, methodical approach to engineering design projects; this purpose is accomplished by presenting project design techniques and then illustrating their use with two major projects. The first project, a clock design, shows how to plan and implement a project. The second project, a complete 68000 microcomputer system, illustrates how to plan a major project; its implementation is described in the Appendix.

Planning and scheduling are vitally important aspects of an engineering project. Finishing the design and prototyping in a reasonable time requires attention to many details. Without proper attention, even the best technically-perfect project can turn out poorly. Consequently, the book stresses how to plan a project and how to schedule a realistic completion date for the project.

WHO SHOULD USE THIS BOOK

This book is written especially for electrical engineering students who are planning and designing projects. Although the project guidelines and illustrations involve digital design and microcomputers, the material can be used effectively by students in other engineering disciplines.

The material in this book is particularly important for students because their future depends on being able to solve the technological problems of society effectively. To get problem-solving experience, students need to practice design by doing projects throughout their years of education. To make the most of design projects, guidelines on problem solving and project planning and implementation can be most helpful. The guidelines in this book provide an organized method of doing a project from start to finish, and they help integrate theory with an understanding of ''real world'' engineering.

HOW TO USE THIS BOOK

The book is intended for use as a supplement to regular course material at all levels of the electrical engineering curriculum. It can be used to provide guidance for introductory-level projects as well as for senior-level capstone design courses. The benefit of the book comes from actively using it in projects of all types; a light reading without application provides no design practice and is not recommended.

Two major design projects are presented in the book. The first one, a clock circuit, is covered from initial planning through finished product. The second, a 68000 microcomputer system, is described and planned; the actual results obtained by a student who built the final product are described in Appendix C. Parts of either or both of these projects may be built and used in an undergraduate project course emphasizing design.

HOW THIS BOOK IS ORGANIZED

Engineering Design: Project Guidelines begins with problem solving and then develops project planning, design techniques, and documentation standards. It illustrates these guidelines with a detailed comprehensive clock design. Next, the requirements for a complete 68000 microcomputer system are presented; its project plan shows how the design can be successfully completed to meet specifications. An actual implementation is given in Appendix C.

Chapter 1 introduces the concept of designing to meet customer needs. Engineering design involves two steps in meeting these needs: first, the project must be defined and planned; second, the project must be implemented. Problem solving is introduced as a tool to help identify the customer requirements as well as to help in all phases of the design process. The interplay of problem solving and project planning is essential to assemble the ''mini-proposal'' at the end of the chapter. The mini-proposal is the guiding document that contains the project definition, objectives, strategy, and step-by-step plan of action with a time schedule. The plan given in the mini-proposal is then used in Chapter 2 to complete a full project implementation. This implementation, beginning with the analysis of the specifications and constraints, covers the technical design of the product on paper and the construction of a working prototype.

Chapter 3 provides the rules and guidelines on how to do a technically sound digital design. Signal levels, loading, timing, and noise are covered in detail because of their importance in building a successful digital system. The material is presented to supplement, not replace, the information contained in an introductory course in digital logic design. A number of heuristics, or rules of thumb, are also included to moderate the technical paper design with a measure of common-sense reality.

Chapter 4 covers the essentials of recording the project in a lab notebook and of documenting the final design. Drawing guidelines for documentation are covered in Chapter 4; standards for the schematic diagrams are given in Appendix A. Rather than describe a "project report" or an in-house technical report, Chapter 4 uses the outline of a "technical manual" as the final system documentation. The reason for the technical manual is simple: far too many excellent designs have become absolutely useless without proper operating and service information. Lack of a comprehensive technical manual can spell disaster in the marketplace. Appendix B contains a sample technical manual that illustrates the ideas presented in Chapter 4.

The clock example in Chapter 5 illustrates the application of project planning and implementation to an actual design project. The secondary purpose of the chapter is to introduce the IEEE Std-696 bus and explain how to use it, or any standard, in a product design.

Chapter 6 provides an overview of a complete 68000 microcomputer system and a description of its requirements. Its purpose is to show how to plan for necessary hardware and software development. This project plan defines the scope of the project and its objectives, states the strategy to reach the objectives, and sketches a plan of action to finally build and test the product. The exercises at the end of the chapter are intended to give direction to the product design described in the Appendix C Technical Manual.*

ACKNOWLEDGMENTS

This book is the result of much help from a number of my students at Bucknell. I would like to thank several in particular: Donna Kidd, Debbie Polstein, Jim Beneke, and Mark Luders. They put in many hours of editing and proofreading the original manuscript used in classes. I appreciate their enthusiastic support throughout this project. Kanwalinder Singh, a graduate student in electrical engineering, participated in my 68000 design course and wrote the technical manual in Appendix C; it is an example of student excellence, and I thank him for his permission to use it here. I also thank Diane Goodling, my assistant at the word processor, for typing the manuscript.

A.D.W.

*The 68000 project is presented completely in *68000 Microcomputer Systems: Designing and Troubleshooting* (Englewood Cliffs, NJ: Prentice-Hall, Inc., 1987).

ONE

Engineering Design
A Creative Activity that Requires Planning

Digital design can be fun! How else can you enjoy building a piece of complex equipment and expect to have it working in several days? If the entire circuit has not already been described in some magazine or book, you can always take parts of the circuit from several sources and piece together a complete design ready to build. Give it a quick ''smoke test'' to see if you have wired it correctly, and then connect it to your home computer. Moreover, after a few quick patches, you can type a program from your favorite magazine and run it in a matter of hours. Give it the ''big bang'' test to see if you have put all the code together correctly, and then congratulate yourself on a successful project.

Do you see a bit of yourself in this? We all experience this scenario for many design projects, and the approach seems to work. It has survived the test of time to become an almost traditional way of developing new hardware and software products. Although not the most efficient way of doing a project, it does appear to get the job done.

Getting the job done is certainly important, but have you really done a proper engineering design? You might have been creative and solved the problem but not practiced sound engineering design along the way. The ''product'' is one of a kind and probably unsuitable for another person to build or for a company to manufacture.

Whether you are a student about to start a major design project or a professional engineer on the job, this hobbyist technique is clearly not satisfactory. You need a systematic way to approach engineering design so that you can complete your project on time and within your budget; in addition, your design must meet all specifications. In short, you must plan your project and plan it well.

1.1 DESIGN OVERVIEW

Engineering design is the creative process of identifying needs and then devising a product to fill those needs. As shown in Figure 1.1, engineering design is the central activity in

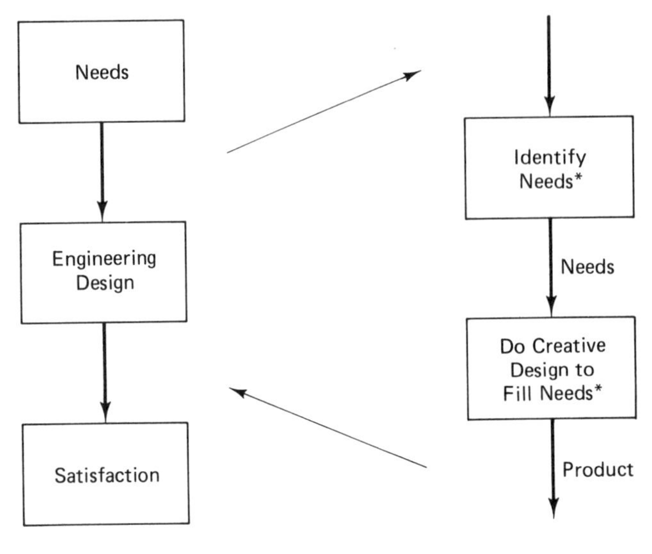

Figure 1.1 Engineering design is the central activity in meeting needs. It involves identifying the needs and then doing a creative design to fill them; both require problem-solving techniques.

meeting needs; these needs may be yours or a customer's. If you understand the requirements involved, then you can develop a creative design to satisfy them.

Figure 1.2 shows two parts of the creative design process. The first part of the process is a project plan outlining the various needs and reducing them to a set of specifications. The project plan is an administrative tool used to identify the various tasks and when to do them. The second part, the project implementation, is the process of designing and developing the final product. Both the project plan and the project implementation are necessary for an orderly product development.

In the context of engineering design, the project plan leads to a set of specifications and tasks. In a sense, you can consider it a nontechnical document because it includes more concepts than technical detail. However, it should not be overlooked. The project plan may be easily summarized and put into the form of a proposal, which is an outline of intended work for a complete project. The proposal functions as a road map for the entire creative design effort, making the difference between project success and failure.

The project implementation, on the other hand, involves the technical activity you would expect in a design project: specifications, hardware and software design and development, documentation, prototype construction, and testing. You can see that the hobbyist technique is only a small part of this implementation and consequently overlooks many essential aspects of the project. Because of these many details, the next chapter is devoted to doing the project implementation.

Both parts of the creative design process require problem solving. It is a problem to determine the information that you need to set the design specifications. Likewise, it is an equally substantial problem to design the product. Both can be addressed by the same problem-solving techniques.

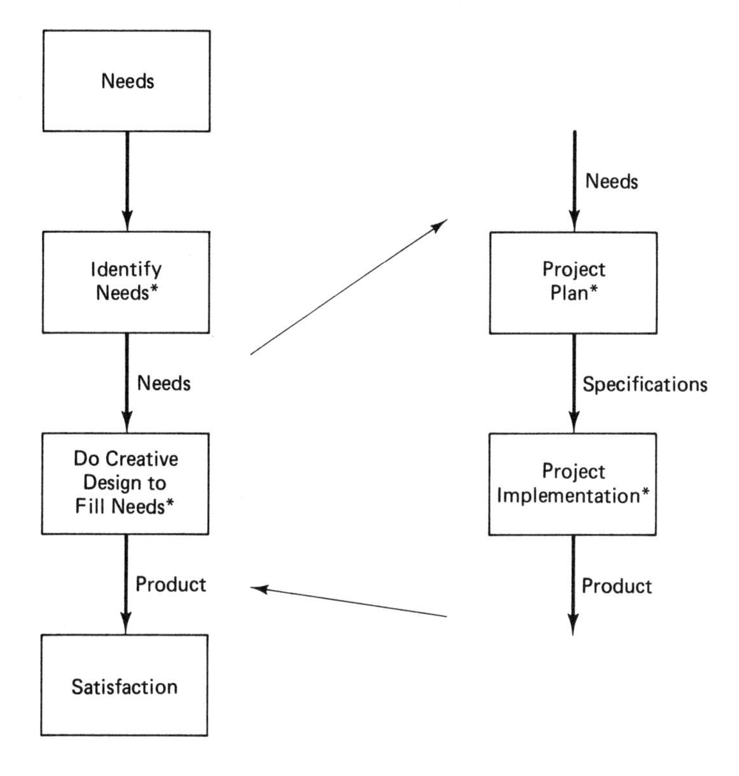

Figure 1.2 The essential first part of a creative design is to complete the project plan. The project plan producess the specifications that describe what product is needed so it can be designed and built.

1.2 PROBLEM SOLVING

Problem solving is the process of determining the best possible action to take in a given situation. This process requires identification of the problem and a description of its causes. Then it makes a systematic evaluation of various alternative solutions until one can be selected as the best. Although you have used problem-solving techniques in one form or another for years, you probably have not looked at them closely.

An outline of a problem-solving method suitable for engineering design is shown in Figure 1.3. It is important to note that this method is not limited to identifying the needs of a customer. It can be used in working through both the project plan and the project implementation. The general problem-solving activities of analysis, synthesis, evaluation, decision making, and action are the essence of engineering design.

Assume for now that you have a customer with a particular technical difficulty. Knowing this customer and his (or her) needs is essential to defining the problem. Who is

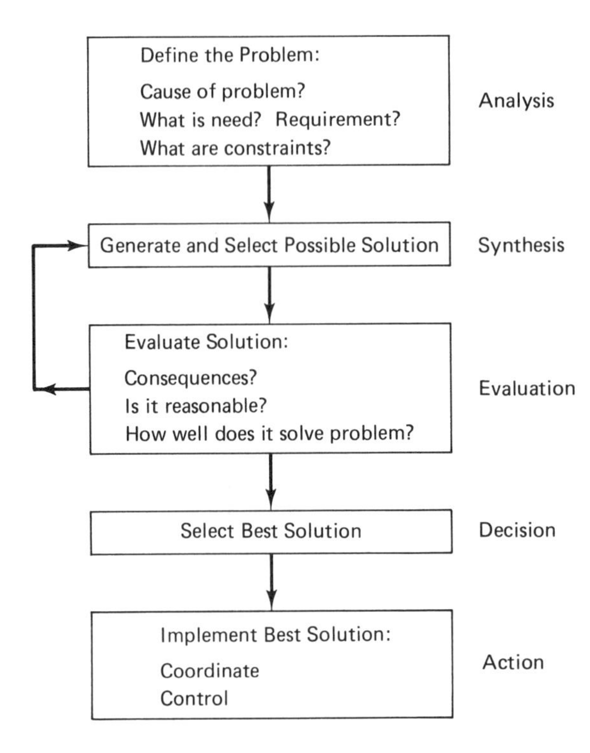

Figure 1.3 General problem-solving steps in defining a problem and evaluating a number of possibilities until a best solution can be selected. The best solution is never perfectly satisfactory because it is a balance between needs and constraints.

the customer? What does he do? What is most important to him in his work? What is least important? What caused his problem? When seeking to define the problem better, be sure to separate the causes of the problem from its effects.

Suppose, for example, that the owner of a local metal-working shop has asked you for your advice on buying a computer. Would you immediately tell him that he needs a Brand-X computer? Of course not. You would begin to ask questions to determine what he wants it to do. Maybe he wants to simplify his job scheduling and inventory management so that he has more free time to plan his future business. Maybe he wants to put all of his accounting on a computer so that he can have quick monthly reports. Maybe a friend has told him that a computer will help business, and "everybody needs a computer to be competitive." What you are doing is analyzing the situation and defining the problem. Remember, if you do not grasp the problem, then any solution will do. This is simply another way of saying, "If you don't know where you're going, then any road will take you there."

How do you find the information necessary to know the customer and his needs? Ask him! If he thinks you can help solve his current problem, he will be more than willing to tell you every problem his company ever had. Understanding his day-to-day operations is vital to defining the problem and its cause.

By the time you understand the problem and its cause, you are also likely to have some ideas on solutions. Make a list of possible solutions, select one, and evaluate its effectiveness. What consequences would you expect from this solution? Any decision you make will have both favorable and unfavorable consequences.

For example, if you were to select Brand-Y computer to solve an accounting problem, you might find these consequences:

1. Computer hardware price reasonable.
2. Software price not too high.
3. This computer might be discontinued soon.
4. Software might do the accounting now.
5. Software might not do for a growing business.

Some consequences look good, while some seem to argue against Brand-Y. What do you do about a list of consequences like this? Set it aside for now and analyze Brand-Z and any other appropriate brands. Perhaps you might compare the brands by constructing a chart or developing a set of standard test programs.

While selecting and evaluating possible solutions to the problem, notice that you are gaining a deeper understanding of the problem. You are also reaching for solutions that you never thought of when you started. You are using facts and concepts to synthesize new ideas. In other words, you are being creative.

Finally, after gathering information and comparing various alternatives, you are ready to make a decision. Any decision, however, involves compromise. After comparing various computer brands, you may find two equally satisfactory choices. What do you do? How do you quantify your preference for the color of Brand-X, or the shape of Brand-Y? The final decision becomes a feeling or preference, an intangible that you cannot define.

After making a decision, you must take action. As in Figure 1.3, the best solution is implemented, coordinated, and controlled. You accomplish this through project management. In the simple computer-selection example, if you were asked for advice on which computer to buy, then your job is done when you give the advice; no real action or project work is involved. On the other hand, if you were asked not only to make a selection, but also to purchase, install, and service the equipment, then you would have a project to implement. This project would require proper planning and close supervision to ensure its success.

1.3 PROJECT PLANNING

A project is a single job that can be accomplished within a specified time and within a certain budget. How this project actually gets done depends on your project plan. The project plan outlines the various needs and reduces them to a set of specifications. It also helps you to identify and schedule the various tasks.

What does this idea of a project plan mean to you, the student or engineer, as you begin designing a piece of hardware or programming a computer system? First, it means that you have an orderly way to conceptualize the confusing array of information. Second, it means that you have an orderly way to complete the project. The project plan is the management tool that helps you do your job.

Look at the project-planning steps outlined in Figure 1.4. How can this outline help you do a better job? Instead of thinking of the project deadline as next year, think of it as next week. Go through the steps of the outline and make a list of the things you should do each of the next five days so that your "one-week" project is a success.

The first step in Figure 1.4 is to define the project. This is a statement of the goals you are trying to accomplish. Think of it as the big picture describing your project. For example, suppose you are to design and build a microcomputer-based temperature monitor. You might write your project definition:

> My project is to design, build, and test a meter that I can use to measure and record air temperature.

At this point, you are saying nothing about its performance (such as temperature range or accuracy); nor are you saying how you intend to build it (do you really need a microcomputer?). Avoid locking yourself into a project goal that is too specific; stick to the big picture.

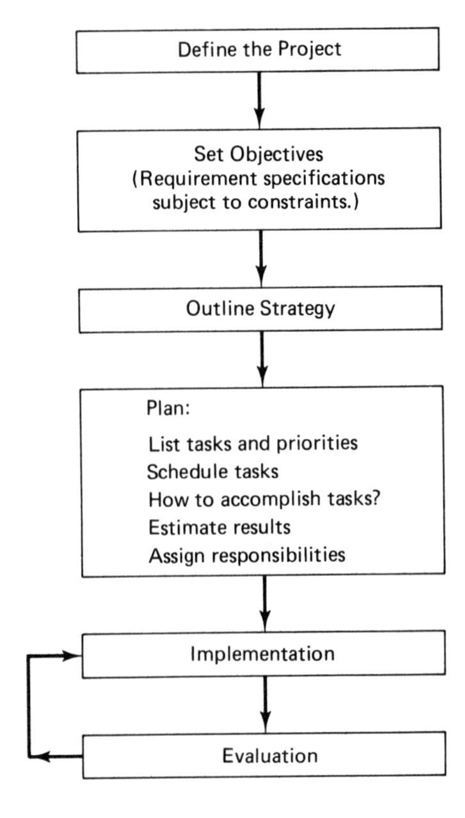

Figure 1.4 General outline of steps to follow when planning a project.

You get specific in the next step when you set your objectives. Your objectives should be specific measurable outcomes of your work. This is where you get into details about the performance of the device you are designing. The objectives are stated in several dimensions: required performance, time to complete, and total cost. For example, you might set these objectives:

> By the end of this month, my meter will be completely built and tested. It will perform to these specifications:
>
> Temperature range −40 to +100°C,
> Accurate to within 1°C,
> Display either Fahrenheit or centrigrade temperature,
> Display minimum and maximum temperatures during last 24 hours,
> Calculate and display 24-hour average temperature,
> Calculate and display daily heating degree days.
>
> In addition to these performance requirements, the meter will be portable and capable of battery operation. Parts for the prototype will cost less than $150.

After you have some solid objectives to work toward, you need to outline your strategy; that is, your concept of how to reach your objectives. Keep in mind that your strategy is your idea of how to achieve the objectives, not the details for actually achieving them. To reach the temperature meter objectives, you might form this strategy:

> To attain my objectives, I will breadboard a prototype model of the analog circuitry with the temperature sensor on the breadboard. Once I understand how it should work, then I will add an analog-to-digital converter plus an interface to a small microcomputer board. I should be able to handle all the calculations and display functions with the microcomputer. After I have it working properly, I will make a prototype printed-circuit board for the customer to evaluate.

This strategy is unique to you and your choice of a design solution. Relate this to what you just learned about problem solving. Your problem is to meet the required design objectives. A possible solution is to start with the breadboard, develop a working circuit, and then build a circuit board. Another possible solution might be to do a complete design on paper first, breadboard it, and then interface it to the microcomputer. Which approach you choose as a best selection depends entirely on you and your work habits. Thus, your strategy is a statement of how you intend to implement the best solution.

You implement your solution by using a plan. Depending on the size of the project, you may find the plan ranging from a simple list of tasks to a complex set of schedules involving many different engineers and departments within your school or company. For our purposes here, we will assume that the implementation is going to involve only one person: yourself.

The best place to start with your plan is to look closely at your strategy and list the major tasks. Early in the project you may not know all your tasks, but you can begin by listing the major tasks chronologically. Under each major task, you probably will think of some subtasks that are also necessary. Some of the major tasks will be high-priority items and must be listed. For example, you must have a breadboard and you must test the circuit for proper operation. Consider this as a possible plan to implement your strategy:

1. Get a breadboard and power supply for the prototype.
2. Look for articles and designs on temperature measurement.
3. Select temperature sensor and A/D for first design.
4. Sketch tentative circuit and calculate circuit values.
5. Build the analog circuit and take measurements.
6. Connect the analog circuit to the A/D converter.
7. Test the circuit completely for proper performance.
8. Design the microcomputer interface logic.
9. Connect the microcomputer and test the interface.
10. Write a simple program to read the temperature.
11. More programs and tasks I cannot estimate now.

Can you start work with this list of tasks? Probably you can. However, you might want to consider how long each will take to complete. If you want to finish the project in a month, when must you complete each task? If you spend the first two weeks on only a few tasks, then you will probably not finish the project in time. You need to set your own deadline for each task.

One way of scheduling is to estimate how many days each task will take and when you should be done with each. As you work your schedule, though, you may lose track if you get ahead or fall behind schedule. You may want to use a bar chart graphical presentation of your schedule as shown in Table 1.1.

TABLE 1.1 BAR-CHART SCHEDULE OF TASKS NEEDED TO BUILD A SIMPLE TEMPERATURE METER

	Week Beginning			
	March			
Tasks to Do	3	10	17	24
Get breadboard, etc.	**			
Get articles	***			
Select sensor and A/D	****			
Sketch circuit	*****			
Build analog and test	*****	****		
Connect analog and A/D	*	*******		
Test completely		******		
Design interface		****		
Connect microcomputer		*	***	
Write simple program			*******	
Write more programs			*****	****
Unknown extra tasks				*******

The bar chart is one of the easiest ways to manage your work schedule as you progress through your project. The chart shows not only the tasks you listed when you started your plan, but also when you will actually start and finish each one. The chart also shows overlapping tasks at a glance. One task might depend on another, as shown in Table 1.1 by the arrow between designing the interface and connecting the microcomputer; when you see the arrow, you know that you must finish one task before starting the other. You can make the chart as simple or as complex as you wish, but remember that you must work from it. It should be usable and easy to modify if you get ahead or behind.

The "unknown extra tasks" designation at the end of the chart reminds you that the chart must be flexible and will be modified as you go along. At the start, you do not know of all the things you need to do or how long each will take. The best you can do is estimate.

As you complete your schedule, think about how you will accomplish each of the tasks. What results do you expect as you go along? Anticipate problems and act as soon as possible. For example, if you think you might need precision components for the accuracy specified, then complete that part of the design early and order the parts so that they will be available when you need them in the circuit. Plan ahead! Also, think of contingency plans so that your project is not in jeopardy if, for example, your precision parts are unavailable.

Once you have a plan and a tentative schedule put together, implement it. Get to work! The implementation part of Figure 1.4 goes hand in hand with evaluation. Because the evaluation is done as you work on your project, you know how closely you are following your plan. Are your results meeting your objectives? Are you getting ahead or behind schedule? You should rethink and modify your schedule to reflect changes. If your project is in trouble, have you asked for help?

Your project manager will want to know how well you are keeping to your schedule and whether you need help with any problems. Normally, you would update management

TABLE 1.2 SAMPLE PROGRESS REPORT COVERING THE SECOND WEEK OF THE TEMPERATURE-METER PROJECT

PROGRESS REPORT Temperature Project—Week 2	
Current status:	The analog design has been completed and successfully tested. There have been no delays and I am on schedule.
Work completed:	During the week since the last report, I completed building and testing the analog circuit. I used the temperature sensor and measured the output of its amplifier and plotted a graph of its response. I connected the A/D converter and tested its performance by varying the temperature sensor voltage.
Current work:	During the last day of this week I started work on the interface design and I am now in the middle of connecting it to the microcomputer board.
Future work:	During the third week I plan to finish the connection to the microcomputer board and to write a program to test the A/D. Then I plan to write a more complex program to display the temperature in both centigrade and Fahrenheit.

MINI-PROPOSAL
Temperature Monitor

Project definition:	The goal of this project is to design, build, and test a meter that can be used to measure and record air temperature.
Project objectives:	At the end of four weeks, the temperature monitor will be completely built and tested. It will perform to these specifications: Temperature range −40 to +100°C Accurate to within 1°C Display either Fahrenheit or centigrade temperature Display minimum and maximum temperatures during last 24 hours Calculate and display 24-hour average temperature Calculate and display heating degree days In addition to these performance requirements, the meter will be portable and capable of battery operation. Parts for the prototype will cost less than $150.
Strategy to achieve objectives:	The analog circuitry and temperature sensor will be prototyped on a temporary breadboard until its operation is fully understood. An analog-to-digital converter plus interface circuit will be added to work with a microcomputer system. After the data is being properly read by the computer, a number of display and calculation programs will be written.
Plan of action:	The various tasks needed to implement the strategy are as follows: Get prototype breadboard and power supply Look for articles and designs on temperature measurement Select temperature sensor and A/D Sketch tentative circuit and calculate circuit values Build analog circuit and take measurements Connect analog circuit to the A/D converter Test the circuit completely Design the microcomputer interface logic Connect microcomputer and test interface Write simple program to read temperature Programs and tasks I cannot estimate now The schedule necessary to finish the project in the required four weeks is attached. [Refer to Table 1.1 for schedule]
Reporting:	Weekly progress reports will be made. At the end of the project a complete engineering design and working prototype will be presented.
Budget:	Initial funding of $150 is necessary to purchase the prototype analog parts and the microcomputer.
Evaluation:	Verification of how well the prototype meets the design specifications subject to the constraints will be made weekly and at the end of the project. The final evaluation will be conducted by the design engineer and the customer.

with a monthly progress report. If you are in school, your professor may require several progress reports during the semester. A progress report can take many forms depending on individual preferences, company policy, or customer requirements. In its simplest form, a progress report describes the current status of your project, the work completed, the work in progress, and your plans leading up to the next report. You should attach a copy of your schedule and mark your progress in each task. A sample progress report is shown in Table 1.2.

As you near the end of your project, you should be evaluating how well you are meeting all of your objectives. Review your progress reports and your schedule. Decide what last-minute action is necessary to correct any difficulties. Leave enough time at the end of the schedule to summarize your project and report on your technical accomplishments.

You can easily summarize all the steps of your project plan in the form of the "mini-proposal" shown in the box. It defines your proposed project, what you want to achieve, and how you plan on doing it. This mini-proposal is abbreviated and illustrates only the major topics you should include in a full proposal. If you are a student, you might find this proposal very useful to focus your project and win financial support to build it. If you are a professional engineer, the proposal is necessary to describe a project to a potential customer. Likewise, a proposal is valuable to gain support for possible areas of new-product development.

1.4 SUMMARY

Computer hardware and software can be easily designed and patched together to work. Usually this hobbyist technique seems to get the job done, but the "design" is probably unsuitable for another person to build. You need a systematic way to approach engineering design so that your product meets specifications and manufacturing requirements.

This systematic approach to engineering design requires that you understand whose needs are involved. Will the product meet your need or your customer's need? Engineering design is the creative process of devising a product to fill customer needs; it closes the gap between customer needs and customer satisfaction. You must use problem-solving skills to find the customer's needs and to design the product properly. If you fail to design the product with your customer in mind, it may work perfectly but be completely useless.

The problem-solving approach is applicable to many situations in addition to the case of determining the customer needs. The steps of defining the problem, selecting a possible solution, evaluating the solution, generating another possible solution, and selecting the best one of many possibilities can be used anywhere. The general problem-solving activities of analysis, synthesis, evaluation, decision making, and action constitute the essence of engineering design.

Problem-solving skills are necessary if you are to know what to do, what decision to make, what road to take, what product to build. But how you carry out a decision requires an understanding of how to plan a project. You must define the project clearly and set objectives. You must devise a strategy and plan your action so that you can complete the project in a reasonable time. You summarize all this in the proposal, in which you describe the various design tasks that go into the final product.

EXERCISES

1. An old family friend just graduated from law school and started working at a local law office. The two attorneys at the firm asked your friend to find a way to automate the typing of legal documents. Using their electric typewriter for all letters and drafts had become increasingly unsuitable as their workload increased over the last several years. Knowing your interest in computers, your friend called you yesterday and asked if a computer is worth investigating.
   a. Who is the customer?
   b. What does the customer need for satisfaction?
   c. Define the problem.
   d. What constraints are there?
   e. List three possible solutions to the problem.
   f. Make a selection. How would you justify it to the customer?
   g. Who is responsible if your idea proves disastrous? Who is responsible if your idea is a great success?
2. Rather than move into a new house, you and your family are going to refurbish your present home. The heating system has been in need of constant repair every year and probably should be replaced if you plan on keeping the house. The winter heating season begins in about two months. Make realistic assumptions based on your past experience.
   a. Define the problem and constraints.
   b. List three possible solutions to the problem and the factors you must consider in making a decision.
   c. Which solution would you choose? Why?
   d. List the tasks needed to implement your solution.
3. Make the temperature-monitor mini-proposal into a proposal to create a device to monitor the local 115 VAC line voltage. There have been a number of complaints that the voltage drops down briefly (how briefly?) when the building's heat pump turns on. This drop is alleged to cause a problem with any computers that are running at the time. You want to find the maximum and minimum voltages as well as the time of day they occurred.
4. Your manager at the Wheel Works just walked into your office to tell you about a new product idea from the marketing department. The idea is to install an electronic water-level indicator on the company's 200-gallon standard steel tank. Besides being able to delete the glass-tube level indicator on the side of the tank, an electronic indicator might even be adapted later to turn on an inlet valve automatically when the water level drops during usage.
   a. Define the problem. Make assumptions about the system.
   b. Make a mini-proposal describing a creative solution.

FURTHER READING

LOVE, SYDNEY F. *Planning and Creating Successful Engineered Designs* (New York: Van Nostrand Reinhold, 1980) (TA174.L68)

MIDDENDORF, WILLIAM H. *Design of Devices and Systems* (New York: Marcel Dekker, 1986) (TA174.M529)

OSTROFSKY, BENJAMIN. *Design, Planning, and Development Methodology*. Englewood Cliffs, NJ: Prentice-Hall, Inc., 1977. (TA174.087)

RAY, MARTYN S. *Elements of Engineering Design: An Integrated Approach*. Englewood Cliffs, NJ: Prentice-Hall, Inc., 1985. (TA174.R37)

ROBERTSHAW, JOSEPH E., STEPHEN J. MECCA, and MARK N. RERICK. *Problem Solving: a Systems Approach*. New York: Petrocelli, 1978. (QA 402.R6.)

RUBINSTEIN, MOSHE F. *Patterns of Problem Solving*. Englewood Cliffs, NJ: Prentice-Hall, 1975.

RUBINSTEIN, MOSHE F., and KENNETH PFEIFFER. *Concepts in Problem Solving*. Englewood Cliffs, NJ: Prentice-Hall, 1980.

WICKELGREN, WAYNE A. *How to Solve Problems*. San Francisco: Freeman, 1974.

TWO

Project Implementation
Bring Ideas to Reality

When you examined engineering design in the last chapter you saw how your needs or your customer's needs could be satisfied. First you used problem-solving techniques to identify needs. Then, after you identified the needs, you prepared a project plan that you developed and summarized in the form of a mini-proposal. This proposal outlined all the necessary work and completion dates for each task.

This chapter shows how to use the proposal to complete the project implementation step shown in Figure 2.1. This step is important because it is a systematic way of finishing the project design within the given time and financial constraints. The approach presented here is just one of many ways to tackle the project implementation, and it may be easily modified to your own particular requirements. The focus is on the method of design rather than on the details of digital circuit design or computer programming.

As indicated in Figure 2.1, the specifications are used in the implementation phase of the engineering design to produce a product meeting your needs or a customer's needs. The project implementation involves two major steps: the technical design and the construction of a prototype. In the technical design step the circuit is created on paper; in the construction phase, the prototype (a working model) is built and tested.

When you look at the temperature monitor mini-project in Chapter 1, you can see the evidence of some technical design work. Although the proposal is an administrative planning document, it contains enough technical effort to establish a set of reasonable specifications and a realistic work schedule. Most of this design was conceptual, and unless you had built something similar in the past, you did not do enough detailed design to be absolutely certain of the results. Consequently, your strategy might be inadequate. However, the mini-proposal does establish some feasible specifications and does provide a useful working document for the full design effort.

Although the mini-proposal is adequate to develop your small project, a typical industrial or government proposal needs a substantial design content. If your company is

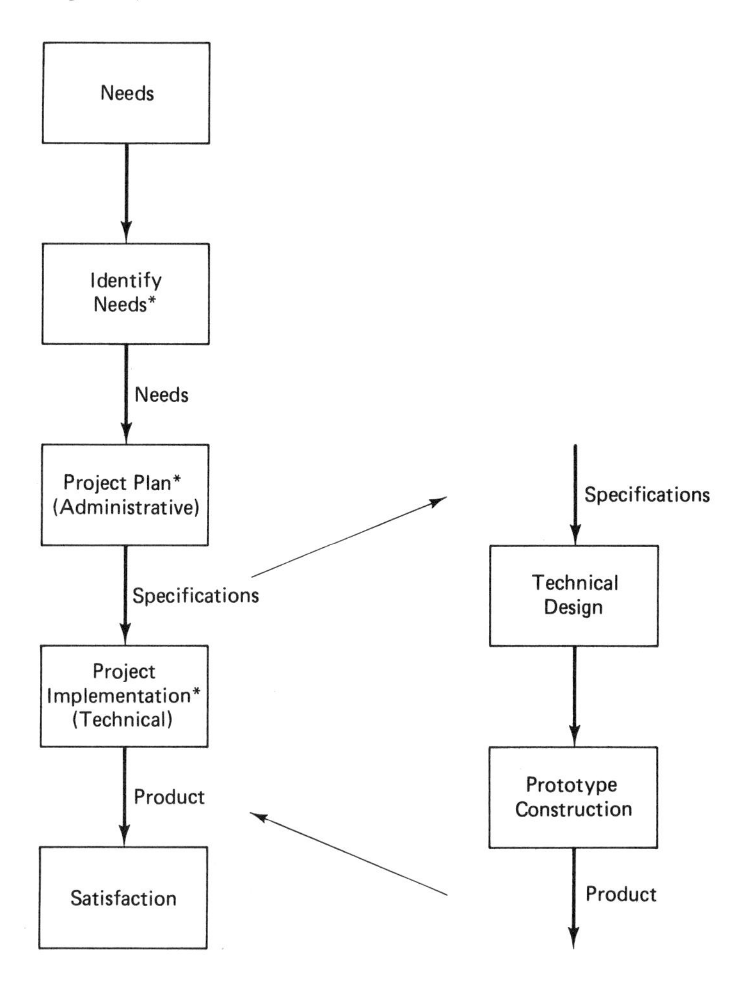

Figure 2.1 Engineering design involves identifying the needs of either you or your customer. Using problem-solving techniques, you can develop a project plan describing the specifications of the needed product. Then, using these specifications, you can implement a project to build the product.

competing for a manufacturing contract, most of the design work, including the completed prototype, will probably be done before the proposal is finalized. You must be certain of your design and be able to estimate closely how much time will be needed to deliver a final product. If your company hopes to make a profit, there is little latitude for errors and radical design changes once the contract has been signed.

For your purposes in learning engineering design, once you have a documented and tested prototype you will consider your engineering complete. However, a prototype that works and meets specifications is by no means a finished job if you are doing the project in industry. If your project is research-oriented, resulting in patents and further research, the prototype is only the beginning. Likewise, if your project is directed toward production, you have much follow-up work to do. For example, your design documentation will be used by drafting personnel to make assembly drawings and schematics. Also, you will be coordinating your project with other electrical, mechanical, production, test, and quality-control engineers so that the final design can be manufactured successfully.

2.1 PROJECT OVERVIEW

The proposal provides an overall plan for the full project implementation. It establishes the project definition, its objectives, and a strategy for meeting those objectives; then it details a plan of action with a schedule for completion. These activities are shown in the left column of Figure 2.2; the corresponding activities in the project implementation are shown in the right column. The planning to set your project's goal and objectives completed the first step of the project implementation. The strategy you developed to solve the design problem was your design concept. The list of tasks and their completion schedule was your technical design combined with prototype construction. On the surface, you might think that these are almost the same. There is a difference: when you carry out the project, you are *doing* the technical design and the prototype construction rather than *talking* about doing it.

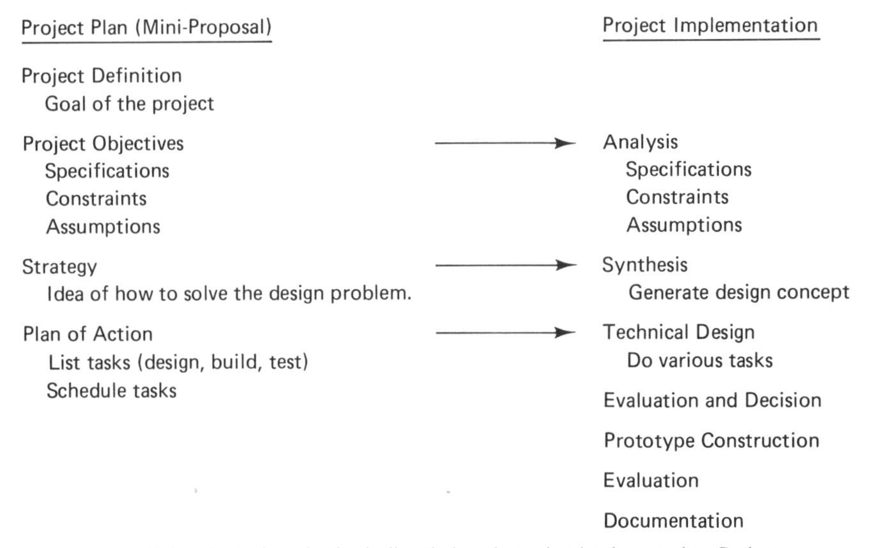

Figure 2.2 The project plan leads directly into the project implementation. Both are closely related throughout the project.

Figure 2.3 shows the overall activities involved in the project implementation. Accomplishing each of these requires a number of steps and may appear somewhat confusing at first. The design sequence is one way of simply visualizing the process you use when designing a product. The major design activities and typical tasks are:

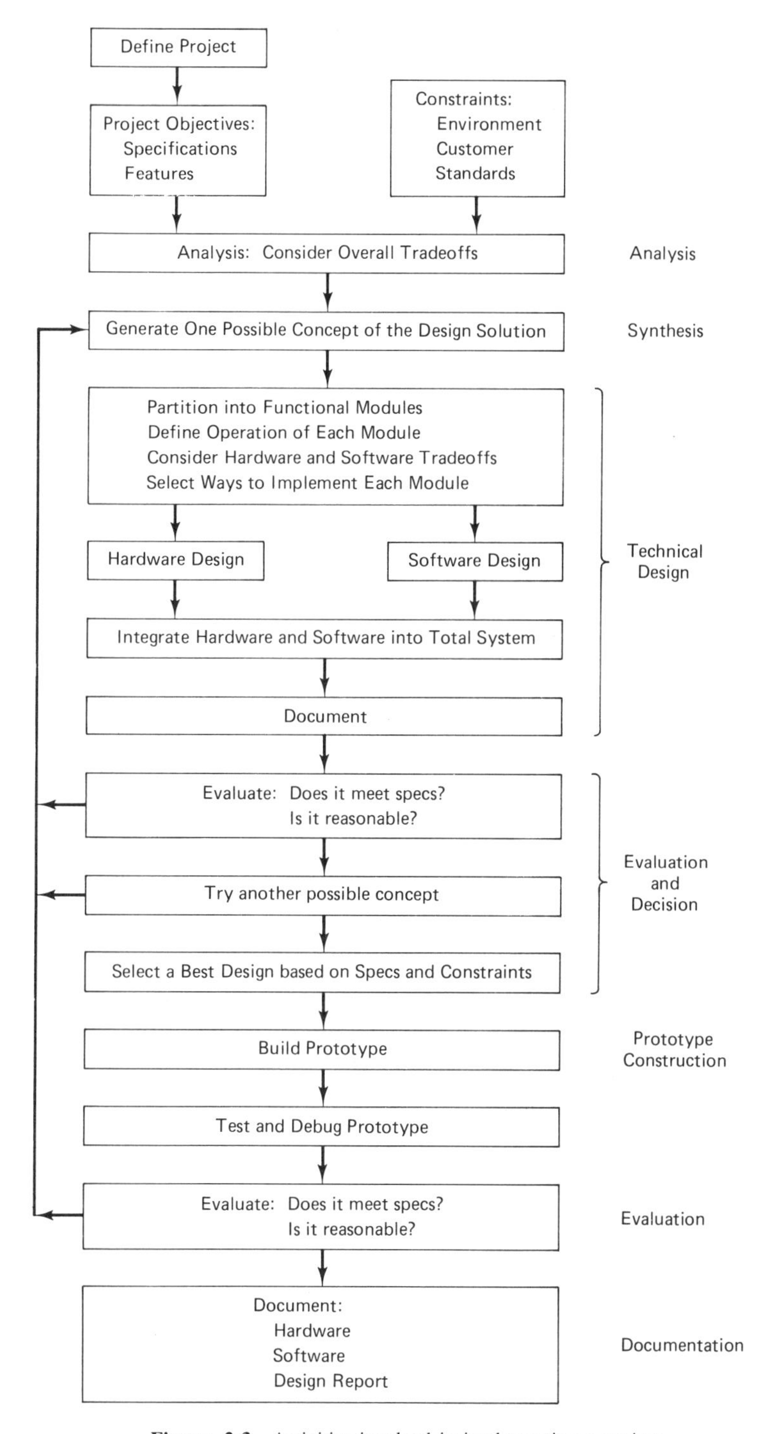

Figure 2.3 Activities involved in implementing a project.

Analysis	Consider the product specifications and features. Allow for constraints related to environment, customer limitations, industry standards. Balance overall tradeoffs between specifications and constraints.
Synthesis	Generate a possible concept of a solution to the design problem subject to the constraints.
Technical Design	Partition system into functional modules and define operation of each. Select ways to implement each module; make tradeoff between hardware and software. Do the circuit design and computer programming, integrate into the system, and document the design so far.
Evaluation/Decision	Review the design. Does it meet the specifications? Is the design reasonable? Try another design and compare the two. Repeat until you can select a best compromise that meets specifications subject to the constraints.
Construction of Prototype	Build a prototype system. Test it and correct any hardware and software errors.
Evaluation	Review the system as built. Does it perform as designed? Does it meet specifications? Is it a reasonable solution to the problem?
Documentation	Gather design documentation and prepare a complete engineering design report.

These design activities are based on the work in the proposal, which by now should be complete. Although they are a systematic way to deal with project implementation, they should not be taken too rigorously: consider this morphology as a guide to design rather than as an inflexible absolute. In all probability, you will do a little of each step, skipping some steps until later, and even going back to the very beginning on occasion to rethink the problem with new insight.

2.2 ANALYSIS

You are analyzing when you break down a system into its component parts and examine each to see how it fits into the system. In this context, you are breaking down the problem statement, the specifications, and the constraints. Then you look closely at each of them to see that it fits well enough to solve the problem.

Begin your design analysis by studying the problem statement as given in the proposal. Investigate the background that led to the problem, and then paraphrase the problem in a sentence or two to ensure that you fully understand it. Review the product features and see if they are consistent with each other and with the specifications. Step back a moment, look at the total situation, and imagine that you have the product already completed as specified. Would it solve the problem? If so, is the solution reasonable? Does it make sense?

The specifications you accept at the start of the project will be your criteria for selecting among the design alternatives. Because the specifications are a statement of your design objectives, they must be as specific as possible so that you know when your design is good enough to start building it. Overengineering a product is perhaps as bad as underengineering: the product never gets built. As you examine the specifications, identify the top-priority requirements and be prepared to consider them first as you solve the design problem.

Resolve any problems with the specifications. Ideally, the specifications should describe the product exactly, but often a product might be overspecified. Trying to meet unnecessary specifications adds extra engineering and complexity to the final product. Similarly, specifications can be ambiguous. Ambiguity results when the writer uses poorly defined or imprecise terms. Specifications can also contradict each other, so that a design solution meeting one requirement would never meet the other.

Review each of the constraints or limits imposed on the product. Are they necessary and realistic, or could some of these limitations be relaxed? Find out how much room you have to work in before you get too involved. For example, if one of your constraints is that the product must be battery-operated and run for 72 hours at full load, find out whether 48 hours would solve the problem. Why? Because the extra battery life might add to the product cost, complexity, and design time. If you know which constraints are flexible, then you can ask for relaxation of certain limitations if necessary.

Some constraints, however, are absolutes. For example, various IEEE standards have been agreed upon concerning certain aspects of electrical design. If you are designing a product that must meet IEEE Std-696, then your design cannot deviate from the limits written in the standard. If your design does not meet the standard on all points, then your product might not work properly with other units in the system. If that happens, then the customer will be inclined to blame any system malfunctions on your product even though your violation might have been inconsequential.

Although standards can at times be an inconvenience, they do make the design job easier because you have a ready-made design outline before you even start. For example, IEEE Std-696 describes the physical and electrical specifications of a circuit board intended for use in a computer system. If you know that your design must conform to this standard, then you immediately know what physical space and what voltages are available for your circuit: the environment for your design will be an enclosure with a maximum of 22 connected devices. As you study the standard and begin to visualize your creation, you can work more easily toward a tentative design solution.

In addition to the explicit constraints, you are working within the implied constraints of a set schedule and limited financial resources. These implied constraints are likely to affect your design decisions substantially. If you had all the time you wanted to complete a design, it would be a masterpiece. If only you had some extra finances, or the customer were willing to pay more, you could do a truly wonderful work of art. Think of the time/money limit as part of the engineering challenge: you would like to design the right product at a reasonable price in time to be useful.

With a set of workable specifications and standards, consider the tradeoffs that you should make between them. When you analyze the specifications, you consider whether each specification is consistent with the next; you consider each of the constraints in a similar manner. You also should investigate an overall tradeoff between specifications and constraints. Consider the battery-life constraint: perhaps you find that the 72-hour life is a ''must,'' but that you can reduce product cost by modifying the specification in another area. The longer battery life might be required, but nothing is said in the specifications about a trickle-charge being used to recharge the batteries constantly before ''battery-only'' operation. If you can keep the batteries fully charged, then perhaps they can be the same size as the 48-hour-life batteries and still run for 72 hours. You might not be able to make the tradeoff even though the specification is ''quiet'' about recharging; on the other hand, you might be able to negotiate a specification requiring charging before use. It depends on the application and all the other variables involved.

The analysis of specifications is never complete. After completing all the steps to implement a project, you will find that another iteration of the analysis improves your understanding. Even though you understood the problem, the objectives, and the constraints, another look can add substantially to the success of your work. In design, you never fully know the situation.

2.3 SYNTHESIS

In the synthesis portion of implementing your project, you are seeking to create and define one concept of the problem solution. Your initial strategy in the proposal described one concept that you believed would work, and from there you developed a project plan. After a complete analysis, you might refine that concept to synthesize a better approach.

As you conclude your analysis, you will also have a number of other ideas for solving the problem. Write them all down even though some of them seem extreme or impractical. Sketching each of your ideas helps you to visualize a potential solution to a design problem. Sort through this assortment of design ideas and pick one that you think will best meet the specifications. This idea might very well be the concept that you use for your final product design; however, you may yet discover a better one. At this point, you make your selection expecting to do a technical design and then an evaluation. You might do several designs before final selection, and each time you go through a design you will have better ideas on how to solve the original problem. This iteration helps refine your concepts into better designs.

Using the concept that best fits the specifications, begin by sketching a block diagram of the system. You can do the block diagram easily by asking what the total system does and drawing one big block with an input and an output. Then look inside the block and draw several small blocks that describe how to use the input to create the output. This is a top-down design approach similar to the software concept of writing a program by starting with the top level module followed by writing its support modules. Consider the temperature monitor from the last chapter where you saw a mini-proposal containing specifications and a plan for design. For the monitor, draw one big box with temperature as an input and display as an output. Then, to obtain the smaller blocks inside, look at the mini-proposal strategy. The strategy (concept) includes a sensor, an analog-to-digital (A/D) converter, an interface, and a microcomputer. You can draw the block diagram of this system as shown in Figure 2.4 to express your strategy for solving the measurement problem.

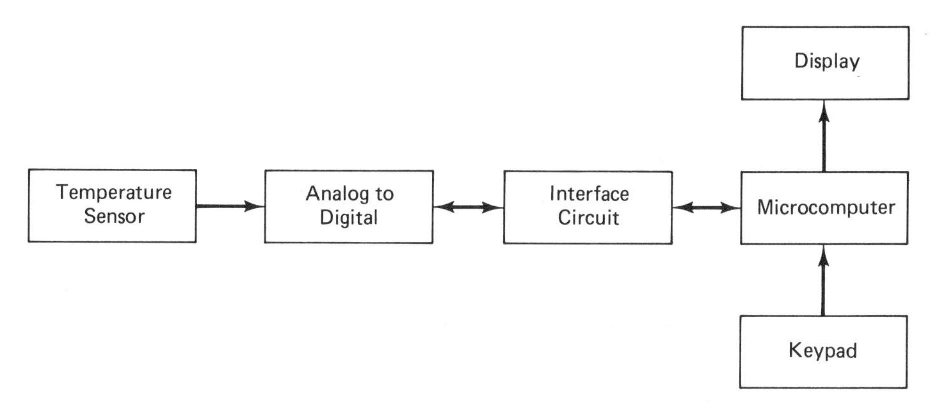

Figure 2.4 Block diagram of the temperature-measurement system. This is a result of partitioning into functional modules.

2.4 TECHNICAL DESIGN

The technical design phase begins with the block diagram of the desired system and finishes with the hardware and software description of the product. The design is on paper in sufficient detail to fully predict how well the product will meet the specifications. For this reason, all your circuit designs and computer programs must be well-documented.

As you draw each of the blocks in the system, you are in effect partitioning the system into functional modules. Each module has a purpose, and you should define each according to its operation, inputs, and outputs. For example, the purpose of the A/D block is to convert an analog voltage to an equivalent digital value. To operate properly, it must be told when to do a conversion, and it must be able to notify the computer when it finishes.

When you partition the system into modules, you actually imply a tradeoff between hardware and software. According to the product concept, the temperature monitor will acquire data and pass it without modification to the computer. If any corrections must be made to compensate for nonlinearities in a thermocouple, for example, the solution must implement them in software. Another concept for solving the measurement problem might have done the compensation in hardware before the A/D converter. Consider the various tradeoffs as you review the design.

The next step in the design is selecting a way to implement each module. Draw a block diagram for each of them to the same level of detail as shown in Figure 2.5 for the

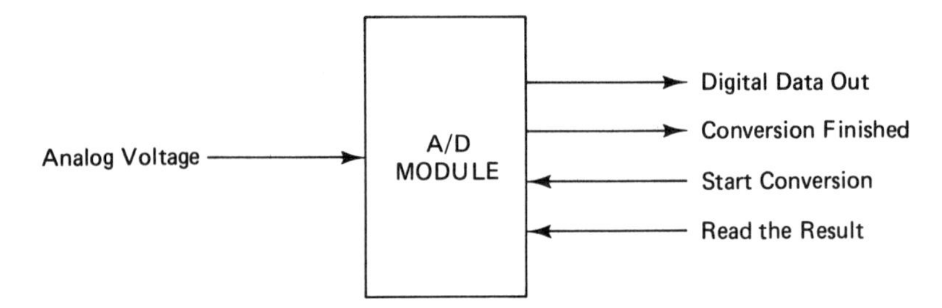

Figure 2.5 The analog-to-digital (A/D) module. This sketch expresses more detail than the overall block diagram of the system shown in Figure 2.4.

A/D module. Then for each module, see if you did a similar design in the past. Can you use your previous work in your current design? Likewise, see if the design has been published in a magazine, book, or manufacturer's application note. For example, how would you implement the A/D converter module? You could build it using transistors and op-amps or perhaps use a single LSI device for the whole A/D function. Suppose you find an LSI chip that meets your accuracy and response-time requirements. In this case you might sketch a rough circuit design like Figure 2.6 to use for the A/D module.

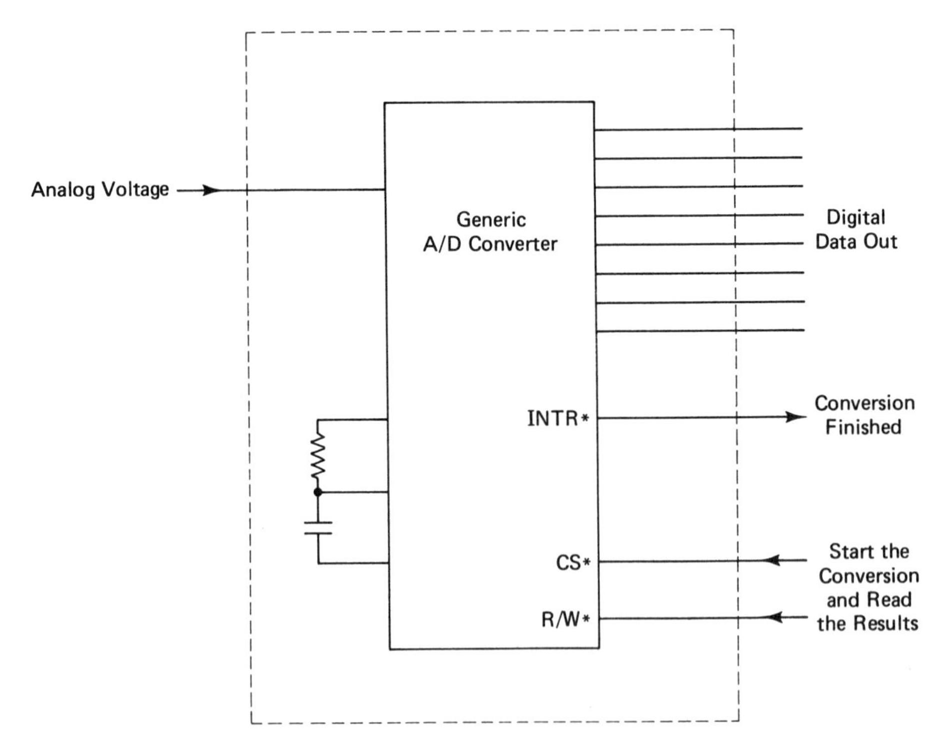

Figure 2.6 An LSI implementation of the A/D Module. This is a rough circuit that needs additional design effort to be operational.

Once you have all the modules roughly sketched, then you can design the detailed circuit. The hardware design at this point is completed to the final circuit with all the components and their interconnections shown in detail. Actual parts should be selected and fully documented. The design should be in accordance with any design rules that have been set for the project. The software design is also done in like detail from the top-level functions down to flow charts and code if possible.

As you design the hardware and software, watch that they parallel one another in their development. Although shown in Figure 2.3 as separate activities, hardware and software design are done concurrently if possible. This is because they must work together when both are integrated into the final system.

Documentation is vital to the success of the hardware and software design. Throughout the analysis, synthesis, and technical design stages of your project implementation you should be maintaining a laboratory notebook. The lab book serves as a permanent reference document of your ideas, plans, and designs as you work through the project. As you make design decisions, you should record the rationale for each decision in the lab book. Also, include in the lab book all the information to make design decisions and to formalize a final design.

2.5 EVALUATION AND DECISION

The evaluation following the technical design stage is intended to see if the hardware and software will perform together as a system and meet the specifications. The evaluation should also reveal what the expected performance should be if the system were actually built.

Suppose you have recently finished a complete paper design of the temperature monitor using the functional modules illustrated in Figure 2.4. You would like to know if your design will meet or exceed the required specifications given in the mini-proposal at the end of Chapter 1. Consider, for example, the requirement for accuracy to within 1° C. To find out how well you meet this accuracy specification, you get the worst-case performance data on the temperature sensor and A/D converter from the manufacturer's data sheets. Also, ask yourself where other errors could come into the system, estimate their size, and add the errors. Your answer will not be exact, but it will be useful. Did it come out to an overall accuracy of 1° or several degrees? If it came out too high, you know you have more design work ahead to improve accuracy.

For each of the product specifications, you might want to make a small chart like Figure 2.7 to help you compare the performance of each design. Include cost, development time, and other possible requirements in addition to the specifications. After you have done one or two designs, you will sense what to expect for a final product and whether or not the original requirements were reasonable for the price. You might find that no design can be completed within your budget in the time available. What do you do then? Perhaps you need to rethink your approach entirely: delete the microcomputer and implement the logic with LSI components. Perhaps you need another meeting with your

	Design 1	*Design 2*
Specification		
Range −40° to +100°C	−40° to +90°C	−40° to +100°C
Accuracy within 1°C	2°	1°
Display Fahrenheit and centigrade	F only	Both OK
Min/Max temperature reading	yes	yes
Average temperature	yes	yes
Heating degree-days	no	yes
Portable	yes	yes
Cost	$60	$200
Time to develop final prototype	3 weeks	6 weeks
Special test equipment required?	no	logic analyzer

Figure 2.7 A simple chart with the mini-project specifications showing how one design compares with another.

customer to review what you can deliver versus the specification that is required; there may be room for negotiation.

The evaluation process goes on continually as you perform the technical design and documentation. As you begin your design, you might find that a particular technical approach cannot meet the specifications. Rather than waste time carrying the design further, estimate its performance, put it in the comparison table, and work up another possible solution to the design problem. Doing this speeds up your technical design substantially, and you have a more diverse selection of alternatives for comparison.

After you complete several designs, your comparison chart will help you decide which design to build as a prototype. The specifications are your decision criteria. Given two designs that meet the specifications equally, you then consider minimum cost, maximum features, maintainability, reliability, and other values important to you or your customer. The actual selection you make, however, should meet the required specifications.

2.6 PROTOTYPE CONSTRUCTION

The purpose of building a prototype is to demonstrate that your paper design is correct and to uncover any oversights that might hinder the product's performance. You can easily use your completed prototype in a number of different experiments to adjust the design and verify its performance.

When you build your prototype, build it module by module. This holds whether you use wire-wrap, a prototype board, or solder for your actual construction. Hardware can be constructed and tested one module at a time just as you might write a computer program one module at a time. The reasoning is this: if you build one small module, apply power, and test it fully for proper operation, then you can use it as part of a larger subsequent module. If any problems develop with the larger module, then you know the difficulty is probably in the circuit you just added. This modular approach to building and testing hardware is far easier than wiring the entire circuit and then trying to diagnose an elusive malfunction in the system.

After you have an operational prototype, test it fully over all the ranges of the specifications. Find and fix any problems that occur as you test the unit. Keep accurate notes in your lab book; it is especially important when you are testing and debugging the prototype. Be able to explain why the unit behaves as it does. Does it perform the way you designed it?

2.7 EVALUATION AND DOCUMENTATION

The evaluation following construction of the prototype is intended to demonstrate that the hardware and software do in fact work together properly and meet the specifications. Unlike the evaluation you did on the prototype, this more formal evaluation will probably involve other persons from different departments in your company. In addition, depending on the product and the arrangements, the customer might receive a copy of the test results.

The results of the prototype testing should prove the merit of your design concept and its implementation. If something was overlooked or a specification changed late in the design cycle, then you might have to redesign even now. Figure 2.3 shows the evaluation going back to start with another concept, but in reality you go back only as far as necessary to correct the problem.

While the prototype is being tested, you should be finishing your final design documentation on the project. Review your lab book and outline your technical design report. Plan on covering the design concept, the reasons for your technical choices, the hardware and software design, the prototype construction, and the operation of the prototype. Besides treating the design details, you should also include a description of how to test the unit and what results should be expected. Later, when your design goes into production, either you or a test engineer will need this test information to write the test plan for the manufactured units.

2.8 SUMMARY

Chapter 2 shows how to finish the project systematically by following a number of engineering design steps in sequence. First, identify needs by using problem-solving techniques, and then prepare a project plan. You can summarize the plan in the form of the mini-proposal. The proposal outlines each task and its completion date and can be used to guide the project implementation. Using the specifications and design concept in the proposal, you can complete the project within the given time and financial constraints.

The activities of analysis, synthesis, technical design, evaluation/design, prototype construction, and evaluation/documentation are presented as a flowchart in Figure 2.3. The sequence is one way of approaching the project implementation and may be easily modified to suit your own needs. View the various activities as guides to aid your design rather than as restraints. You will probably begin each step and then on occasion return to the start to rethink with a new understanding of your project.

When you analyze, you are breaking down the problem statement, the specifications, and the constraints to see how well they fit together to solve the problem. Because your planned design must meet the specifications, resolve any problems with them such as ambiguity, contradiction, and overspecificity; resolve any similar problems with the stated constraints as well. If you are designing to conform to a particular industry standard, be sure that your specifications are consistent with its requirements.

In the synthesis portion of the project implementation, you are attempting to create and define one concept of the problem solution. After considering a number of ideas, pick one concept that appears likely to fulfill your specifications the best. Then sketch a block diagram that expresses this concept.

You use this block diagram in the technical design phase to complete a paper design of your project. Each block in the diagram is a functional module; you can describe each according to its operation, inputs, and outputs. Select how you can implement each module in hardware and software, and then design the circuit and program in detail.

The evaluation following the technical design is intended to see if the hardware and software will perform together as a system that meets the specifications. Make a comparison chart to establish which of your design alternatives is most satisfactory. When you have several possibilities, use the specifications as your criteria to decide which design to build as a prototype.

The purpose of building the prototype is to demonstrate that your paper design is correct and to uncover any oversights that might adversely affect the product's performance. Construct the prototype module by module so that you can test each section as you build it. When you have all the modules interconnected, fully test and debug your finished prototype.

The final evaluation after you debug the prototype is intended to demonstrate that the hardware and software work together as a system and that they meet specifications. The documentation should result in a technical design report during this phase of the project. This report should cover design concept, reasons for your choices, and the design, construction, and operation of the prototype.

EXERCISES

1. During the heating season, fuel-oil distributors make customer deliveries based on how cold the weather has been rather than filling tanks on a weekly or biweekly basis. Needless frequent deliveries increase costs, so it is desirable to wait until the customer tank is nearly empty before filling it. Ideally, a typical 275-gallon tank should be refilled when it gets to within 20 to 50 gallons of being empty.

Hot-Spot Oil company presently estimates a "k-factor" for each customer to help decide when to deliver oil. The k-factor is an empirical number of degree-days of heating the customer gets from each gallon of oil. Heating degree-days equal 65 minus the average temperature during 24 hours; for example, if the average temperature yesterday was 20°, then 45 degree-days of heating were required. Suppose Hot-Spot estimated Jack Smith's house at $k = 5$ degree-days per gallon: the 45 degree-days of heat required yesterday used 45/5, or 9 gallons of oil.

If Hot-Spot Oil can keep data on daily heating degree-days, then they can estimate how much oil Jack Smith is using, If Jack has a 275-gallon tank, then that means he can heat for a maximum of 275 times 5, or 1375 degree-days. If each day averages 20°, then Jack will run out of oil in about 30 days (1375/45 = 30+). To be on the safe side, Hot-Spot will probably deliver about 5 days before they estimate Jack will run out of oil. This is equivalent to about 1100 degree-days accumulated since the last delivery.

Hot-Spot Oil came to you recently, and you both worked out some specifications for a way to calculate yesterday's number of degree-days each morning when they come to work. For Jack and their other customers, they can total the degree-days and figure out when to make deliveries. The specifications you worked out are: measure temperature from −40 to +70°F, calculate the average temperature from 8 A.M. to 8 A.M., calculate the number of degree-days, and display the result for them to write down.

a. Define the problem.

b. List the specifications above and add three more specifications you might want to include for a better definition of the job.

c. List three constraints.

d. List three possible ways to solve the problem.

e. Do a rough sketch of each way you listed.

f. Select the best approach and do a block diagram of the system.

g. Using your block diagram, partition the system by function.

h. Do a detailed block diagram of each module.

i. Do the circuit design for at least one module.

j. Describe how you would build your prototype.

k. Describe an alternative way to construct the prototype.

2. Make up a problem and specify a product that you can design. Do a rough mini-proposal and then all the implementation steps from problem definition through prototype description (steps a–k in Problem 1). Polish the mini-proposal after completing these steps. Your deliverable to the customer is a completed mini-proposal.

Pick an interesting topic from either your own background or the following list of ideas:

Computer-controlled speech synthesizer
Programmable music organ
Waveform generator
ASCII display of serial or parallel data
Joystick or mouse for computer cursor control
Printer buffer
Clock with alarm
Data modem
Morse code generator
Programmable power supply

FURTHER READING

ARTWICK, BRUCE A. *Microcomputer Interfacing*. Englewood Cliffs, NJ: Prentice-Hall, 1980. (TK 7888.3.A86)

Asimow, Morris. *Introduction to Design*. Englewood Cliffs, NJ: Prentice-Hall, 1962. (TA 175.A83)

Cain, William D. *Engineering Product Design*. London: Business Books Limited, 1969. (TA 174.C3)

Comer, David J. *Digital Logic and State Machine Design*. New York: Holt, Rinehart and Winston, 1984. (TK 7868.59C66)

Davis, Thomas W. *Experimentation with Microprocessor Applications*. Reston VA: Reston Publishing, 1981.

Fletcher, William I. *An Engineering Approach to Digital Design*. Englewood Cliffs, NJ: Prentice-Hall, 1980. (TK 7868.D5F5)

Gregory, S. A. *Creativity and Innovation in Engineering*. London: Butterworth, 1972. (TA 174.C7X)

Harman, Thomas L., and Barbara Lawson. *The Motorola MC68000 Microprocessor Family: Assembly Language, Interface Design, and System Design*. Englewood Cliffs, NJ: Prentice-Hall, 1985. (QA 76.8.M6895H37)

Hayes, John P. *Digital System Design and Microprocessors*. New York: McGraw-Hill, 1984. (TK 7874.H393)

Kline, Raymond M. *Structured Digital Design*. Englewood Cliffs, NJ: Prentice-Hall, 1983.

Robertshaw, Joseph E., Stephen J. Mecca, and Mark N. Rerick. *Problem Solving: a Systems Approach*. New York: Petrocelli, 1978. (QA 402.R6)

Short, Kenneth L. *Microprocessors and Programmed Logic*. Englewood Cliffs, NJ: Prentice-Hall, 1981. (QA76.5.S496)

Winkel, David, and Franklin Prosser. *The Art of Digital Design*. Englewood Cliffs, NJ: Prentice-Hall, 1980. (TK 7888.3W56)

THREE

Design Rules and Heuristics

Guidelines and Common Sense

In the first chapter, after you identified needs, you prepared a project plan that you summarized in the form of a mini-proposal. This proposal outlined all the necessary work and completion dates for each task. Then, in the second chapter, you used the proposal to do the project implementation by starting with the specifications in the proposal. First you completed a technical design on paper, and then you built a working prototype; as you did this implementation, you were following the general activities shown in Figure 2.3.

This chapter provides guidelines for doing the technical design on paper before actually building the prototype. These guidelines or design rules are the conventions established to do an orderly design that is not only technically sound but also consistent with the designs of team members working on the same project. In addition to providing technical guidelines, this chapter presents a number of heuristics, or rules of thumb, that may be applied to the design process. These heuristics serve to moderate the technical paper design with a measure of common-sense reality.

The technical design you did in Chapter 2 involved using a block diagram of one particular concept. Before building a prototype, you expected to go through a design of several different concepts, and you hoped that one of them would meet the specifications subject to the constraints. When you did the technical design using your block diagram, you probably did it rather haphazardly and did not concern yourself with following any particular rules: if the numbers worked together, then you were happy enough to finish.

When you follow design rules, however, your work goes easier and is more orderly. For these rules to make sense though, some additional detail beyond the original ideas in Figure 2.3 might be helpful. Figure 3.1 shows how the technical design can be expanded and used to go from a block diagram to a complete paper design. Starting with the concept, you can partition your system into functional modules and describe the purpose and function of each module. At the same time, you can decide on how to split the various functions between hardware and software. Then, for each hardware and software module,

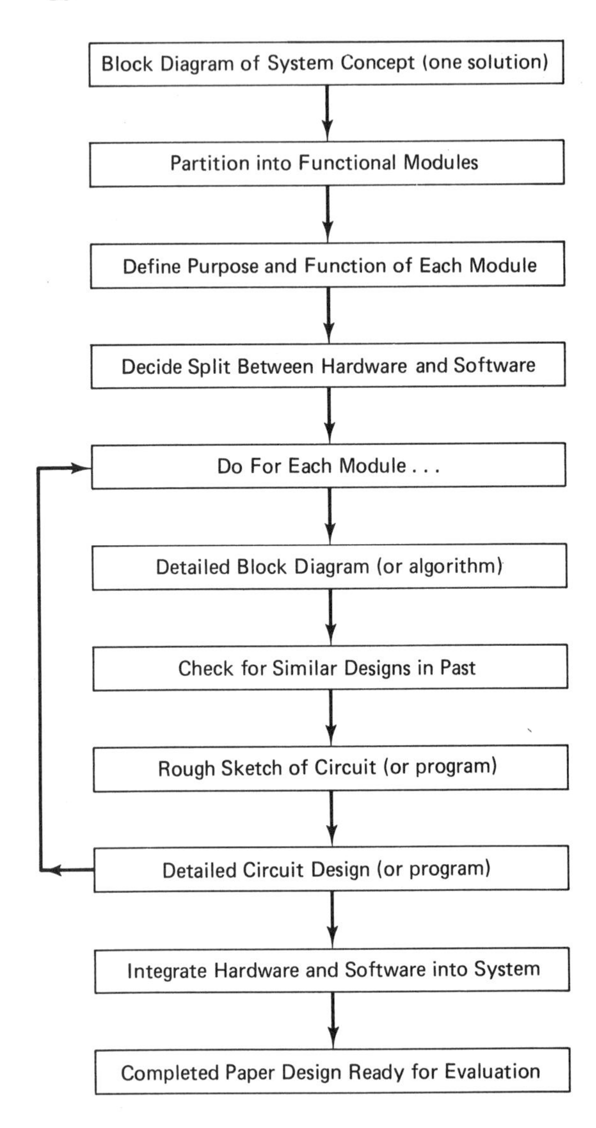

Figure 3.1 An expanded form of the technical design first presented in Figure 2.3

you can draw a block diagram or write an algorithm and do a detailed circuit design or working program. The design rules in this chapter typically apply when you do the detailed circuit design or the program.

Heuristics, on the other hand, apply anywhere in design engineering from the original problem-solving steps through to the detailed circuit design. Heuristics are techniques or rules of thumb that help solve a problem, but are not themselves technically justifiable. They are a blend of past experience, logic, common sense, and nonsense that gives the engineer some direction in solving the problem at hand.

3.1 TECHNICAL DESIGN RULES

After you have a block diagram of your intended circuit, you will probably want to quickly check some of your previous work for similar designs. Many times you can locate a useful circuit in your own notes, in a magazine, or in a text and save some time for extra effort on the more unique parts of your design job. This is not to say, however, that you can simply select a handy circuit and put it in your system without analysis; even if the circuit appears to fit exactly, you should always verify that it does indeed meet your specifications.

Based on your understanding of the circuit requirements, draw a rough sketch of the circuit. At this point, the idea is not necessarily to have a working circuit. You want to get an idea of the ICs involved and how they all fit together in the module. Pinpoint areas you think are difficult or might cause problems later in your design.

Finally, using the rough sketch as a guide, do a detailed circuit design. As you work on the design, you might be unsure how a particular device works; find out—take it to the lab with the data book and try it out. While you do the design work, review the following design rules as a guide. The rules will help you transform the rough circuit sketch into a technically sound design. In addition, if you are working with other designers on a large project, the design rules will help maintain consistency so that the various systems modules will work together.

3.1.1 Hardware Design Rules

Most of the digital logic devices you will use in your design work will be either TTL or CMOS family components. There are a number of advantages and disadvantages associated with each family, and the decision to use one, the other, or a combination of both depends on a number of factors. Generally, if you want speed, then you select TTL; if you want noise-immunity and low-power battery operation, then you select CMOS.

Both TTL and CMOS families simplify the digital design process considerably by allowing the connection of the digital circuit in building-block fashion; that is, a particular logical function can be quickly designed by simply interconnecting the appropriate logic gates to each other. Gates in the same family can be freely connected without concern about how the gates work internally; when using LSI devices, this freedom saves considerable design time.

When gates in the same family are connected, one concern is that proper logic signal levels must be maintained as additional gates are interconnected. Within the TTL family, for example, a logic LOW may be between 0 and 0.8 V and a logic HIGH may be between 2.0 and 5 V. The addition of more and more gates to the output of a single device will tend to cause the LOW to go up and the HIGH to come down. Thus, if the circuit is overloaded, the LOW and HIGH might become 1 V and 1.9 V, respectively; these levels might not be recognized properly as LOW and HIGH later in the system and cause malfunctions.

When gates from different logic families are connected, maintaining the proper signal levels is even more critical. Within the CMOS family, if the devices are connected to a

5-V supply, a logic LOW may be between 0 and 1.5 V, and a logic HIGH may be between 3.5 and 5 V. You can see that the CMOS logic LOW does not match with TTL: if the CMOS device outputs 1.5 V as a LOW, there is a strong likelihood that a TTL input will misinterpret it as a logic HIGH. Usually the TTL and CMOS parts can be interfaced easily, but you must pay close attention to design for the proper signal levels. While TTL devices can easily drive CMOS, the typical CMOS gate does not have the drive capability to connect directly to TTL.

In the context of your design work in this book, TTL parts will be used almost exclusively. Consequently, most of the design rules are related to design in the one family, and very little attention is given to interfacing or using CMOS. If you have a design requirement for CMOS, gather component data and design following the same general principles as you would using TTL.

Timing is another important issue in hardware design. It takes a finite length of time for signals in a digital system to get to their destination, and special care must be taken to consider all aspects of signal propagation. Problems with clock skew, setup times, and hold times should be resolved before the circuit design is finalized.

Synchronous sequential circuits should be used in your system design. Using clocked sequential circuits helps make your design considerably easier because you can draw timing diagrams and relate the state transitions to clock pulses. These timing diagrams can be used in troubleshooting your circuit as well as in design. In general, noise is always a problem, but with a synchronous device, noise present at the input is ignored unless it comes at just the instant the clock pulse arrives.

In contrast, asynchronous sequential circuits operate without a clock: input signals ripple through the system, set and reset flip-flops, and produce an output at some unpredictable future time depending on the system propagation delays. Also, because signals can happen any time in the asynchronous circuit, it is prone to being upset by noise in the system. For example, a noise burst between clock pulses in a synchronous system would be ignored, but in the asynchronous system the noise could cause a number of flip-flops to change state and cause a system malfunction.

Use worst-case specifications for your components when you do hardware design. For example, when you estimate the power requirements for your circuit, add up all the worst-case specified values to obtain an upper bound on your power needs. When you do your timing diagrams, the worst-case propagation delays in your analysis to help estimate circuit speed.

Signal levels and loading. When you examine the data sheets for TTL parts, you find that there are two different logic device outputs available: totem-pole and open-collector. The difference is that the open-collector device does not have an internal pull-up resistor. Compare the outputs of the circuits shown in Figure 3.2. Most of your digital circuits can be done entirely with totem-pole devices; however, if you need to connect the outputs of several devices (as to a bus or in a wired-OR), then you use an open-collector device. Both of these TTL devices have the same input characteristics and both should be treated as logical equivalents when you do a logic diagram.

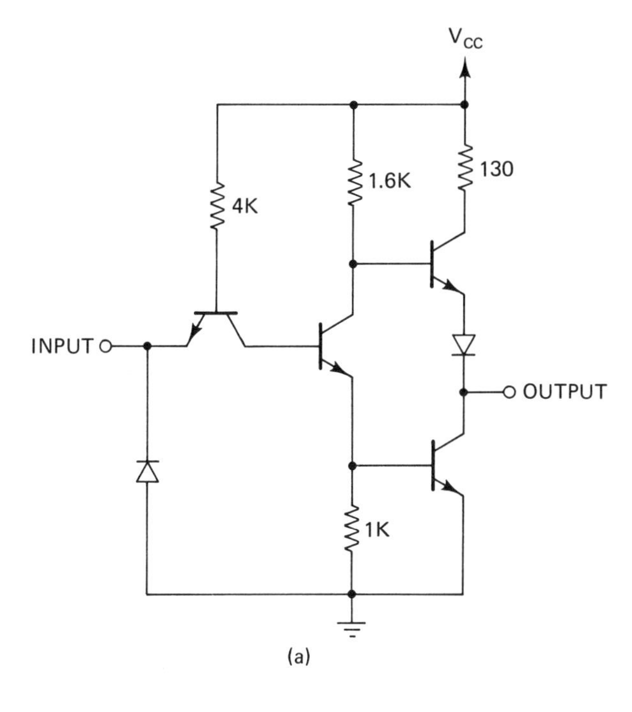

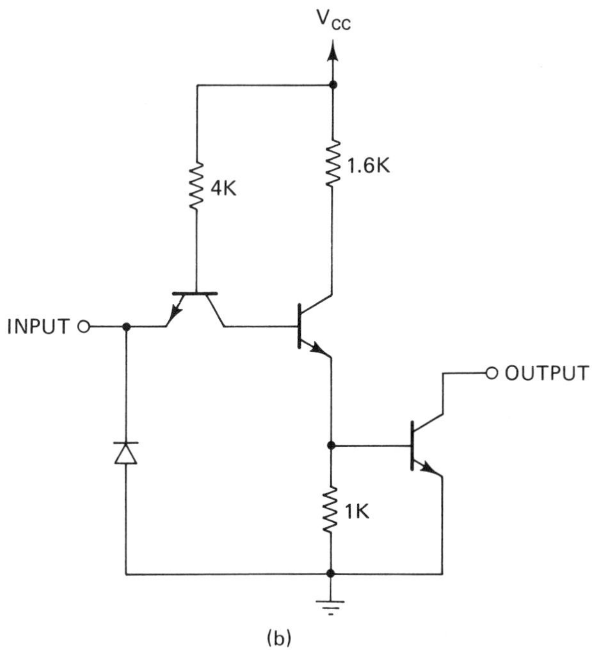

Figure 3.2 Circuits of typical TLL Devices. (a) 7404 totem-pole output, (b) 7405 open-collector output. (Courtesy Texas Instruments, Inc.)

The voltages corresponding to the logic LOW and HIGH are given by the TTL specifications for V_{IL}, V_{IH}, V_{OL}, V_{OH}. These voltages are typically given by

V_{IL} = 0.8 V = the maximum input voltage that will be interpreted as a logic LOW.

V_{IH} = 2.0 V = the minimum input voltage that will be interpreted as a logic HIGH.

V_{OL} = 0.4 V = the maximum logic LOW output voltage.

V_{OH} = 2.4 V = the minimum logic HIGH output voltage.

Visualize a single 7404 inverter gate. Any input voltage from 0 to V_{IL} (0.8 V) will be interpreted as a logic LOW and cause a logic HIGH output; this logic HIGH output can be anything from 5 V down to V_{OH} (2.4 V). Similarly, any input voltage from V_{IH} (2.0 V) up to 5 V will be interpreted as a logic HIGH and cause a logic LOW output; this logic LOW output can range from 0 to a maximum of V_{OL} (0.4 V). The typical output voltages will probably be above 3.4 V for a HIGH and less than 0.2 V for a LOW.

These voltages are illustrated in Figure 3.3. Notice that an input voltage between 0.8 and 2.0 V is not defined; any input voltage in this range will cause an unpredictable output. If the input voltage is below V_{IL} or above V_{IH}, then the device is guaranteed to interpret the input correctly as a LOW or HIGH, respectively. Likewise, an output logic LOW will be below 0.4 V and an output logic HIGH will be above 2.4 V. Any output voltage between 0.4 and 2.4 V is guaranteed not to occur unless the device is overloaded.

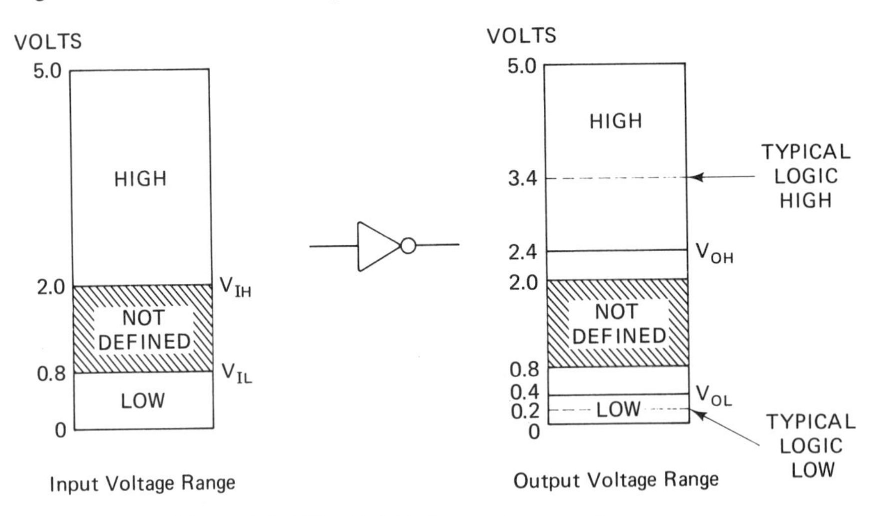

Figure 3.3 TTL allowable input and output voltage ranges.

Look at the data sheet, Figure 3.4, for information on the positive NANDs and inverters. Notice that currents are specified for the inputs and outputs of the gates:

I_{IL} = current flowing out of an input when LOW (V_{IL}) applied.

I_{IH} = current flowing into an input when HIGH (V_{IH}) applied.

recommended operating conditions

		SN5400			SN7400			UNIT
		MIN	NOM	MAX	MIN	NOM	MAX	
V_{CC}	Supply voltage	4.5	5	5.5	4.75	5	5.25	V
V_{IH}	High-level input voltage	2			2			V
V_{IL}	Low-level input voltage			0.8			0.8	V
I_{OH}	High-level output current			– 0.4			– 0.4	mA
I_{OL}	Low-level output current			16			16	mA
T_A	Operating free-air temperature	– 55		125	0		70	°C

electrical characteristics over recommended operating free-air temperature range (unless otherwise noted)

PARAMETER	TEST CONDITIONS †			SN5400			SN7400			UNIT
				MIN	TYP‡	MAX	MIN	TYP‡	MAX	
V_{IK}	V_{CC} = MIN,	I_I = – 12 mA				– 1.5			– 1.5	V
V_{OH}	V_{CC} = MIN,	V_{IL} = 0.8 V,	I_{OH} = – 0.4 mA	2.4	3.4		2.4	3.4		V
V_{OL}	V_{CC} = MIN,	V_{IH} = 2 V,	I_{OL} = 16 mA		0.2	0.4		0.2	0.4	V
I_I	V_{CC} = MAX,	V_I = 5.5 V				1			1	mA
I_{IH}	V_{CC} = MAX,	V_I = 2.4 V				40			40	μA
I_{IL}	V_{CC} = MAX,	V_I = 0.4 V				– 1.6			– 1.6	mA
I_{OS}§	V_{CC} = MAX			– 20		– 55	– 18		– 55	mA
I_{CCH}	V_{CC} = MAX,	V_I = 0 V			4	8		4	8	mA
I_{CCL}	V_{CC} = MAX,	V_I = 4.5 V			12	22		12	22	mA

† For conditions shown as MIN or MAX, use the appropriate value specified under recommended operating conditions.
‡ All typical values are at V_{CC} = 5 V, T_A = 25°C.
§ Not more than one output should be shorted at a time.

switching characteristics, V_{CC} = 5 V, T_A = 25°C (see note 2)

PARAMETER	FROM (INPUT)	TO (OUTPUT)	TEST CONDITIONS		MIN	TYP	MAX	UNIT
t_{PLH}	A or B	Y	R_L = 400 Ω,	C_L = 15 pF		11	22	ns
t_{PHL}						7	15	ns

NOTE 2: See General Information Section for load circuits and voltage waveforms.

POST OFFICE BOX 225012 • DALLAS, TEXAS 75265

Figure 3.4 (Courtesy Texas Instruments, Inc.)

recommended operating conditions

		SN54LS00			SN74LS00			UNIT
		MIN	NOM	MAX	MIN	NOM	MAX	
V_{CC}	Supply voltage	4.5	5	5.5	4.75	5	5.25	V
V_{IH}	High-level input voltage	2			2			V
V_{IL}	Low-level input voltage			0.7			0.8	V
I_{OH}	High-level output current			− 0.4			− 0.4	mA
I_{OL}	Low-level output current			4			8	mA
T_A	Operating free-air temperature	− 55		125	0		70	°C

electrical characteristics over recommended operating free-air temperature range (unless otherwise noted)

PARAMETER	TEST CONDITIONS †			SN54LS00			SN74LS00			UNIT
				MIN	TYP‡	MAX	MIN	TYP‡	MAX	
V_{IK}	V_{CC} = MIN,	I_I = − 18 mA				− 1.5			− 1.5	V
V_{OH}	V_{CC} = MIN,	V_{IL} = MAX,	I_{OH} = − 0.4 mA	2.5	3.4		2.7	3.4		V
V_{OL}	V_{CC} = MIN,	V_{IH} = 2 V,	I_{OL} = 4 mA		0.25	0.4		0.25	0.4	V
	V_{CC} = MIN,	V_{IH} = 2 V,	I_{OL} = 8 mA					0.35	0.5	
I_I	V_{CC} = MAX,	V_I = 7 V				0.1			0.1	mA
I_{IH}	V_{CC} = MAX,	V_I = 2.7 V				20			20	μA
I_{IL}	V_{CC} = MAX,	V_I = 0.4 V				− 0.4			− 0.4	mA
I_{OS}§	V_{CC} = MAX			− 20		− 100	− 20		− 100	mA
I_{CCH}	V_{CC} = MAX,	V_I = 0 V			0.8	1.6		0.8	1.6	mA
I_{CCL}	V_{CC} = MAX,	V_I = 4.5 V			2.4	4.4		2.4	4.4	mA

† For conditions shown as MIN or MAX, use the appropriate value specified under recommended operating conditions.
‡ All typical values are at V_{CC} = 5 V, T_A = 25°C
§ Not more than one output should be shorted at a time, and the duration of the short-circuit should not exceed one second.

switching characteristics, V_{CC} = 5 V, T_A = 25°C (see note 2)

PARAMETER	FROM (INPUT)	TO (OUTPUT)	TEST CONDITIONS		MIN	TYP	MAX	UNIT
t_{PLH}	A or B	Y	R_L = 2 kΩ,	C_L = 15 pF		9	15	ns
t_{PHL}						10	15	ns

NOTE 2: See General Information Section for load circuits and voltage waveforms.

Figure 3.4 (continued) (Courtesy Texas Instruments, Inc.)

recommended operating conditions

		SN5404			SN7404			UNIT
		MIN	NOM	MAX	MIN	NOM	MAX	
V_{CC}	Supply voltage	4.5	5	5.5	4.75	5	5.25	V
V_{IH}	High-level input voltage	2			2			V
V_{IL}	Low-level input voltage			0.8			0.8	V
I_{OH}	High-level output current			−0.4			−0.4	mA
I_{OL}	Low-level output current			16			16	mA
T_A	Operating free-air temperature	−55		125	0		70	°C

electrical characteristics over recommended operating free-air temperature range (unless otherwise noted)

PARAMETER	TEST CONDITIONS †	SN5404			SN7404			UNIT
		MIN	TYP‡	MAX	MIN	TYP‡	MAX	
V_{IK}	V_{CC} = MIN, I_I = −12 mA			−1.5			−1.5	V
V_{OH}	V_{CC} = MIN, V_{IL} = 0.8 V, I_{OH} = −0.4 mA	2.4	3.4		2.4	3.4		V
V_{OL}	V_{CC} = MIN, V_{IH} = 2 V, I_{OL} = 16 mA		0.2	0.4		0.2	0.4	V
I_I	V_{CC} = MAX, V_I = 5.5 V			1			1	mA
I_{IH}	V_{CC} = MAX, V_I = 2.4 V			40			40	μA
I_{IL}	V_{CC} = MAX, V_I = 0.4 V			−1.6			−1.6	mA
I_{OS} §	V_{CC} = MAX	−20		−55	−18		−55	mA
I_{CCH}	V_{CC} = MAX, V_I = 0 V		6	12		6	12	mA
I_{CCL}	V_{CC} = MAX, V_I = 4.5 V		18	33		18	33	mA

† For conditions shown as MIN or MAX, use the appropriate value specified under recommended operating conditions.
‡ All typical values are at V_{CC} = 5 V, T_A = 25°C.
§ Not more than one output should be shorted at a time.

switching characteristics, V_{CC} = 5 V, T_A = 25°C (see note 2)

PARAMETER	FROM (INPUT)	TO (OUTPUT)	TEST CONDITIONS	MIN	TYP	MAX	UNIT
t_{PLH}	A	Y	R_L = 400 Ω, C_L = 15 pF		12	22	ns
t_{PHL}					8	15	ns

NOTE 2: See General Information Section for load circuits and voltage waveforms.

POST OFFICE BOX 225012 • DALLAS, TEXAS 75265

Figure 3.4 (continued) (Courtesy Texas Instruments, Inc.)

TYPES SN54LS04, SN74LS04
HEX INVERTERS

recommended operating conditions

		SN54LS04			SN74LS04			UNIT
		MIN	NOM	MAX	MIN	NOM	MAX	
V_{CC}	Supply voltage	4.5	5	5.5	4.75	5	5.25	V
V_{IH}	High-level input voltage	2			2			V
V_{IL}	Low-level input voltage			0.7			0.8	V
I_{OH}	High-level output current			– 0.4			– 0.4	mA
I_{OL}	Low-level output current			4			8	mA
T_A	Operating free-air temperature	– 55		125	0		70	°C

electrical characteristics over recommended operating free-air temperature range (unless otherwise noted)

PARAMETER	TEST CONDITIONS †	SN54LS04			SN74LS04			UNIT
		MIN	TYP ‡	MAX	MIN	TYP ‡	MAX	
V_{IK}	V_{CC} = MIN, I_I = – 18 mA			– 1.5			– 1.5	V
V_{OH}	V_{CC} = MIN, V_{IL} = MAX, I_{OH} = – 0.4 mA	2.5	3.4		2.7	3.4		V
V_{OL}	V_{CC} = MIN, V_{IH} = 2 V, I_{OL} = 4 mA		0.25	0.4			0.4	V
	V_{CC} = MIN, V_{IH} = 2 V, I_{OL} = 8 mA					0.25	0.5	
I_I	V_{CC} = MAX, V_I = 7 V			0.1			0.1	mA
I_{IH}	V_{CC} = MAX, V_I = 2.7 V			20			20	μA
I_{IL}	V_{CC} = MAX, V_I = 0.4 V			– 0.4			– 0.4	mA
I_{OS} §	V_{CC} = MAX	– 20		– 100	– 20		– 100	mA
I_{CCH}	V_{CC} = MAX, V_I = 0 V		1.2	2.4		1.2	2.4	mA
I_{CCL}	V_{CC} = MAX, V_I = 4.5 V		3.6	6.6		3.6	6.6	mA

† For conditions shown as MIN or MAX, use the appropriate value specified under recommended operating conditions.
‡ All typical values are at V_{CC} = 5 V, T_A = 25°C.
§ Not more than one output should be shorted at a time, and the duration of the short-circuit should not exceed one second.

switching characteristics, V_{CC} = 5 V, T_A = 25°C (see note 2)

PARAMETER	FROM (INPUT)	TO (OUTPUT)	TEST CONDITIONS	MIN	TYP	MAX	UNIT
t_{PLH}	A	Y	R_L = 2 kΩ, C_L = 15 pF		9	15	ns
t_{PHL}					10	15	ns

NOTE 2: See General Information Section for load circuits and voltage waveforms.

POST OFFICE BOX 225012 • DALLAS, TEXAS 75265

Figure 3.4 (continued) (Courtesy Texas Instruments, Inc.)

I_{OL} = current flowing into an output when it is LOW.

I_{OH} = current flowing out of an output when it is HIGH, or leakage current going into a turned-off open-collector output with a specified HIGH output voltage applied.

A sketch with arrows indicating current directions is shown in Figure 3.5 for the 7404 and 74LS04. The spec-sheet treats current as positive if it enters the devices, and negative if it comes out; the arrows in the sketch give the actual direction of the current. If you draw sketches with arrows, then your analysis will make more sense and be easier to understand.

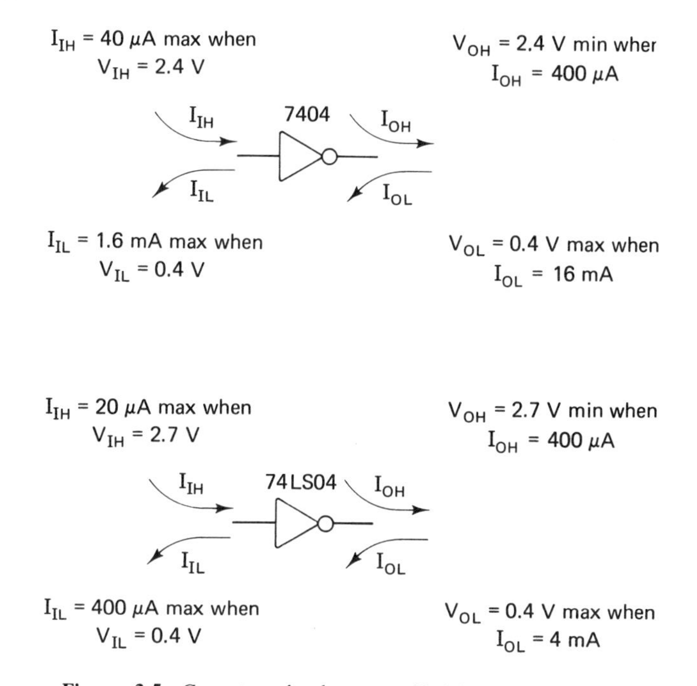

Figure 3.5 Currents and voltages specified for 7404 and 74LS04.

Designing With Totem-Pole Devices. Although it is simple to interconnect totem-pole devices, you must always consider both the currents and the voltages involved. As more gates are connected onto the output of a single device, more current flows and causes the logic HIGH to drop down and the LOW to go up. If too many loads are connected, then the output logic level does not meet the required 0.8 and 2.0 V. Consider these TTL example circuits:

Example 1

Connect two 7404 inverters to the output of a single 7404. Suppose the input to "a" is LOW, so its output is HIGH at a level of at least V_{OH}. The current I from "a" flows into the two inputs, each of which require a maximum of I_{IH}, for a total of 80 μA. This is a light load on "a," and the output voltage V_{OH} stays above the 2.4 V minimum.

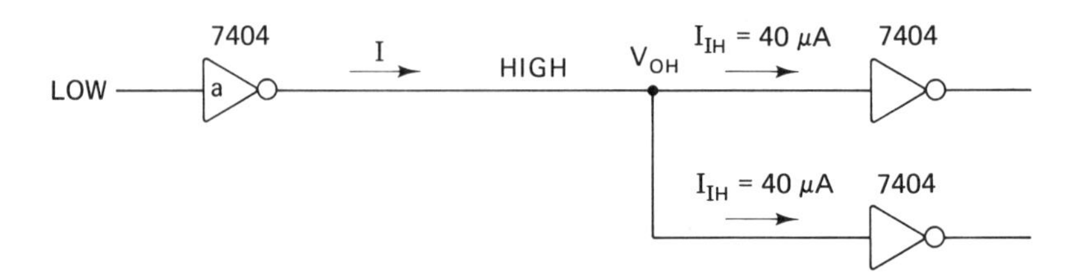

If you connect ten 7404s to the output of "a," then the total current, I_{OH}, required will be 10 × 40 μA or 400 μA maximum. This load tends to pull the output voltage, V_{OH}, down to 2.4 V in the worst case. If more than ten loads are put on the single 7404 gate, then the output voltage will go lower still, and finally go below 2 V; when this happens, the loads will not operate reliably because their input voltage is below their required V_{LH} logic HIGH.

Example 2

Connect two 7404 inverters to the output of a single 7404. Suppose the input to "a" is HIGH, so its output, V_{OL}, is LOW. The current I_{IL} flows from the two inputs, each of which requires a maximum of I_{IL}, for a total of 3.2 mA. This is a light load on "a," and the output voltage V_{OL} stays below the 0.4 V maximum for a LOW.

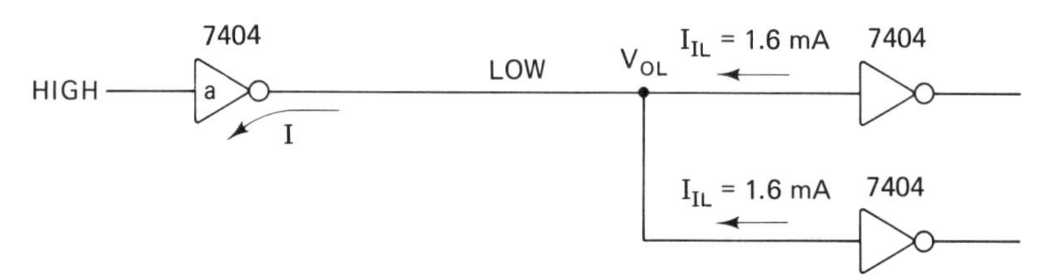

If you connect ten 7404s to the outut of "a," then the total current, I_{OL}, required will be 10 × 1.6 mA, or 16 mA maximum. This load tends to pull the output voltage, V_{OL}, up to 0.4 V in the worst case. If more than ten loads are put on the single 7404 gate, then the output voltage will go higher still, and finally be above 0.8 V; the loads will not operate reliably because their input voltage is above their required V_{IL} logic LOW.

Therefore, you can see that after considering both the HIGH and LOW logic cases, a maximum of ten gates can be connected to a single 7404 inverter. This is referred to as a "fanout" of ten. By following a similar line of reasoning, you can calculate a fanout for each of the logic gates you select for your system.

Designing with Open-Collector Devices. Open-collector TTL components do not have the active pull-up of the totem-pole devices. If you connected a 7405 inverter with an LED on the output as shown in Figure 3.6(a), you would see nothing, even when you might expect an active HIGH. If you changed the circuit slightly plus added an external pull-up resistor as shown in Figure 3.6(b), then the circuit would perform correctly. Consider the output transistor as an on-off switch: when on, the LED is at ground potential and lights; when off, the LED circuit is open and it remains unlighted.

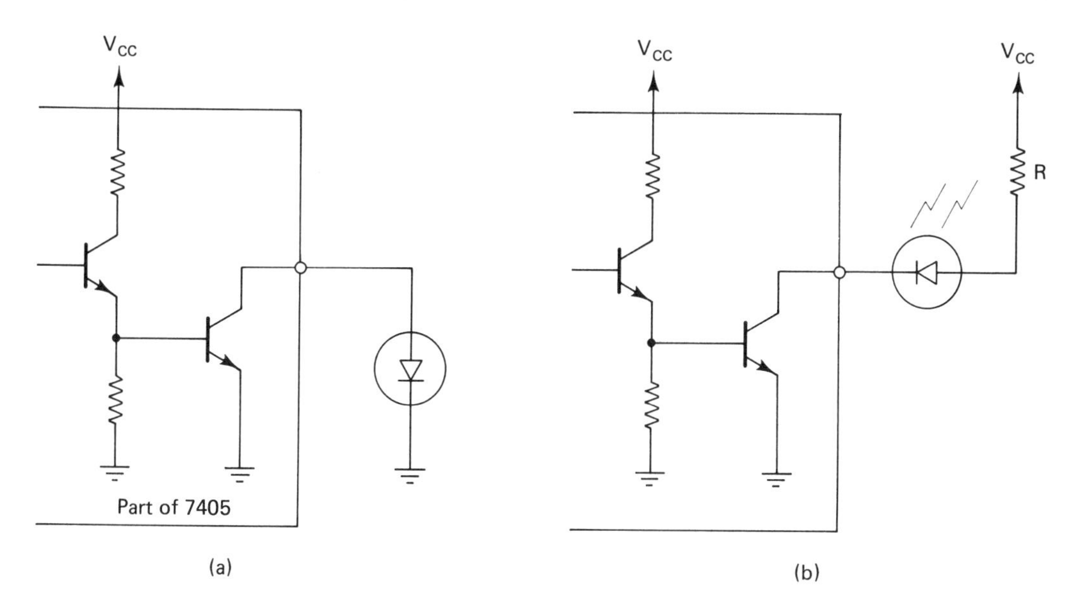

Figure 3.6 A pull-up resistor must be used on the output of the open-collector IC.

The logic levels that correspond to a logic LOW and HIGH are the same as the normal TTL specifications. They are

$$V_{IH} = 2.0 \text{ V} \qquad V_{OH} = 2.4 \text{ V}$$

$$V_{IL} = 0.8 \text{ V} \qquad V_{OH} = 0.4 \text{ V}$$

If you look at the data sheet in Figure 3.7 for the 7405, however, you see that the sign given for the output current is different. This results in the voltages and currents shown in Figure 3.8. Compare this sketch with the totem-pole situation in Figure 3.5.

Instead of being able to source current, the open-collector device has an ouput leakage current I_{OH}, which is defined:

I_{OH} = leakage current going into a turned-off open-collector output with a specified HIGH output voltage applied.

To find a value for the external pull-up resistor, you need to examine the circuit when it is a logic LOW and then examine it again when it is a logic HIGH. When you do this, you find a minimum and maximum value for the resistor. Pick some compromise resistance that allows you maximum response speed but does not draw too much current.

recommended operating conditions

		SN5405			SN7405			UNIT
		MIN	NOM	MAX	MIN	NOM	MAX	
V_{CC}	Supply voltage	4.5	5	5.5	4.75	5	5.25	V
V_{IH}	High-level input voltage	2			2			V
V_{IL}	Low-level input voltage			0.8			0.8	V
V_{OH}	High-level output voltage			5.5			5.5	V
I_{OL}	Low-level output current			16			16	mA
T_A	Operating free-air temperature	– 55		125	0		70	°C

electrical characteristics over recommended operating free-air temperature range (unless otherwise noted)

PARAMETER	TEST CONDITIONS†			MIN	TYP‡	MAX	UNIT
V_{IK}	V_{CC} = MIN,	I_I = – 12 mA				– 1.5	V
I_{OH}	V_{CC} = MIN,	V_{IL} = 0.8 V,	V_{OH} = 5.5 V			0.25	mA
V_{OL}	V_{CC} = MIN,	V_{IH} = 2 V,	I_{OL} = 16 mA		0.2	0.4	V
I_I	V_{CC} = MAX,	V_I = 5.5 V				1	mA
I_{IH}	V_{CC} = MAX,	V_I = 2.4 V				40	μA
I_{IL}	V_{CC} = MAX,	V_I = 0.4 V				– 1.6	mA
I_{CCH}	V_{CC} = MAX,	V_I = 0 V			6	12	mA
I_{CCL}	V_{CC} = MAX,	V_I = 4.5 V			18	33	mA

† For conditions shown as MIN or MAX, use the appropriate value specified under recommended operating conditions.
‡ All typical values are at V_{CC} = 5 V, T_A = 25°C.

switching characteristics, V_{CC} = 5 V, T_A = 25°C (see note 2)

PARAMETER	FROM (INPUT)	TO (OUTPUT)	TEST CONDITIONS		MIN	TYP	MAX	UNIT
t_{PLH}	A	Y	R_L = 4 kΩ,	C_L = 15 pF		40	55	ns
t_{PHL}			R_L = 400 Ω,	C_L = 15 pF		8	15	ns

NOTE 2: See General Information Section for load circuits and voltage waveforms.

POST OFFICE BOX 225012 • DALLAS, TEXAS 75265

Figure 3.7 (Courtesy Texas Instruments, Inc.)

recommended operating conditions

		SN54LS05			SN74LS05			UNIT
		MIN	NOM	MAX	MIN	NOM	MAX	
V_{CC}	Supply voltage	4.5	5	5.5	4.75	5	5.25	V
V_{IH}	High-level input voltage	2			2			V
V_{IL}	Low-level input voltage			0.7			0.8	V
V_{OH}	High-level output voltage			5.5			5.5	V
I_{OL}	Low-level output current			4			8	mA
T_A	Operating free-air temperature	− 55		125	0		70	°C

electrical characteristics over recommended operating free-air temperature range (unless otherwise noted)

PARAMETER	TEST CONDITIONS†	SN54LS05			SN74LS05			UNIT
		MIN	TYP‡	MAX	MIN	TYP‡	MAX	
V_{IK}	V_{CC} = MIN, I_I = − 18 mA			− 1.5			− 1.5	V
I_{OH}	V_{CC} = MIN, V_{IL} = MAX, V_{OH} = 5.5 V			0.1			0.1	mA
V_{OL}	V_{CC} = MIN, V_{IH} = 2 V, I_{OL} = 4 mA		0.25	0.4		0.25	0.4	V
	V_{CC} = MIN, V_{IH} = 2 V, I_{OL} = 8 mA					0.35	0.5	
I_I	V_{CC} = MAX, V_I = 7 V			0.1			0.1	mA
I_{IH}	V_{CC} = MAX, V_I = 2.7 V			20			20	μA
I_{IL}	V_{CC} = MAX, V_I = 0.4 V			− 0.4			− 0.4	mA
I_{CCH}	V_{CC} = MAX, V_I = 0 V		1.2	2.4		1.2	2.4	mA
I_{CCL}	V_{CC} = MAX, V_I = 4.5 V		3.6	6.6		3.6	6.6	mA

† For conditions shown as MIN or MAX, use the appropriate value specified under recommended operating conditions.
‡ All typical values are at V_{CC} = 5 V, T_A = 25°C.

switching characteristics, V_{CC} = 5 V, T_A = 25°C (see note 2)

PARAMETER	FROM (INPUT)	TO (OUTPUT)	TEST CONDITIONS	MIN	TYP	MAX	UNIT
t_{PLH}	A	Y	R_L = 2 kΩ, C_L = 15 pF		17	32	ns
t_{PHL}					15	28	ns

NOTE 2: See General Information Section for load circuits and voltage waveforms.

POST OFFICE BOX 225012 • DALLAS, TEXAS 75265

Figure 3.7 (continued) (Courtesy Texas Instruments, Inc.)

I_{IH} = 40 μA max when V_{IH} = 2.4 V

Leakage I_{OH} = 250 μA max when V_{OH} = 5.5 V

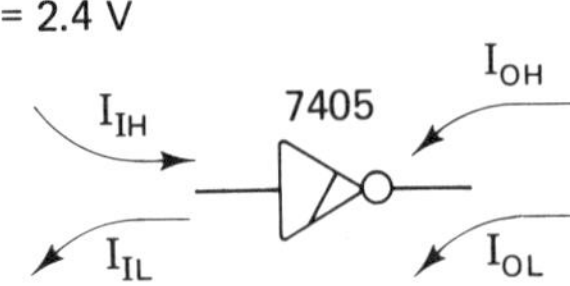

I_{IL} = 1.6 mA max when V_{IL} = 0.4 V

V_{OL} = 0.4 V max when I_{OL} = 16 mA

I_{IH} = 20 μA max when V_{IH} = 2.7 V

Leakage I_{OH} = 100 μA when V_{OH} = 5.5 V

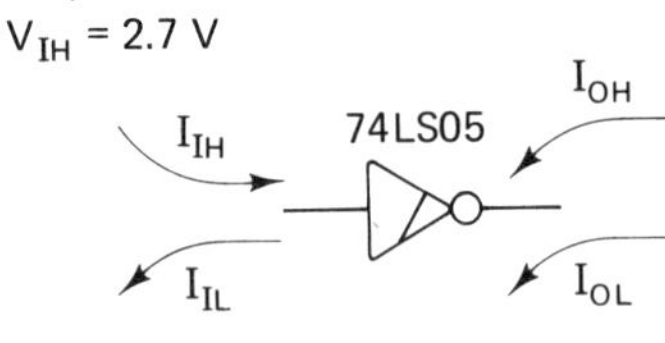

I_{IL} = 400 μA max when V_{IL} = 0.4 V

V_{OL} = 0.4 V max when I_{OL} = 4 mA

Figure 3.8 Currents and voltages specified for the 7405 and 74LS05 open-collector ICs.

Consider these example problems:

Example 3

You are given a circuit with a 7405 open-collector inverter connected to a 7404. What value pull-up resistor should you specify?

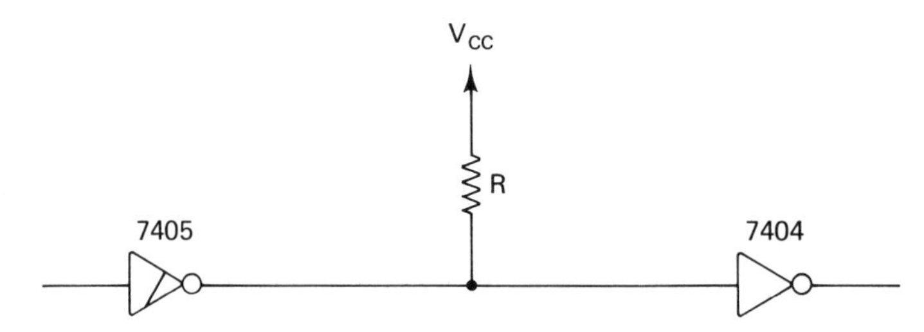

(a) For a logic LOW at the output of the 7405, you would expect currents and voltages:

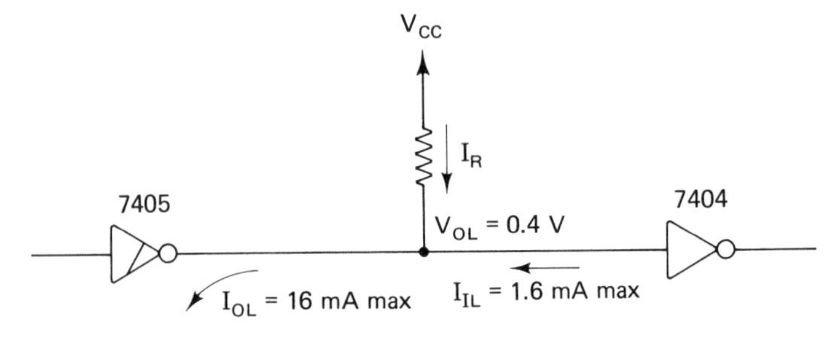

If you design for the maximum I_{OL} into the 7405, then the current through the resistor is

$$I_R = I_{OL} - I_{IL}$$
$$= 16 - 1.6 \text{ mA}$$
$$= 14.4 \text{ mA}$$

With the 7405 sinking 16 mA, its output voltage will rise to a maximum of $V_{OL} = 0.4$ V. The voltage across the resistor is

$$V_R = V_{CC} - V_{OL}$$
$$= 5 - 0.4$$
$$= 4.6 \text{ V}$$

and this makes the minimum resistor value

$$R_{\text{min}} = 4.6 \text{ V}/14.4 \text{ mA}$$
$$\approx 320\ \Omega$$

(b) Do a similar calculation for a logic HIGH at the 7405 output:

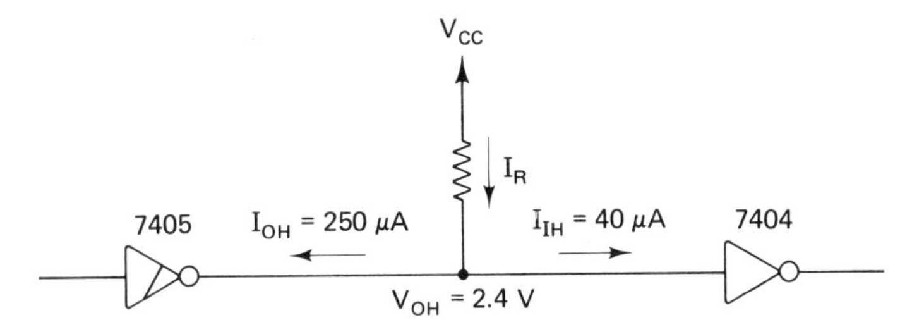

The current through the resistor is

$$I_R = I_{OH} + I_{IH}$$
$$= 250 + 40\ \mu\text{A}$$
$$= 290\ \mu\text{A}$$

and the voltage across the resistor is

$$V_R = V_{CC} - V_{OH}$$
$$= 5 - 2.4$$
$$= 2.6 \text{ V}$$

so the maximum value of the resistor is

$$R_{\text{max}} = 2.6 \text{ V}/290\ \mu\text{A}$$
$$\approx 8970\ \Omega$$

(c) Select some value of resistor between the two extremes. When the resistor is small, you gain better circuit speed at the expense of the added current when the 7405 output is at a logic LOW. For a large resistor, you save current but lose some speed on the rising edge of a pulse from the 7405. You might select a value of $R = 2.2$ KΩ as a possible compromise.

Example 4

Consider a typical situation you might encounter when you decode an address. The circuit partially decodes the address bus using a wired-AND circuit. When the proper address appears on A7, A6, and A5, the output is asserted. Suppose that the switches are set to ''101'' so that any address in the range 101x xxxx (that is, AO through BFhex) will cause a chip-select (CS) response. In the wired-AND, all the 74LS136 outputs must be off (logic HIGH) so the resistor can pull up CS. If just one of the 74LS136 outputs go LOW, then that is sufficient to negate CS.

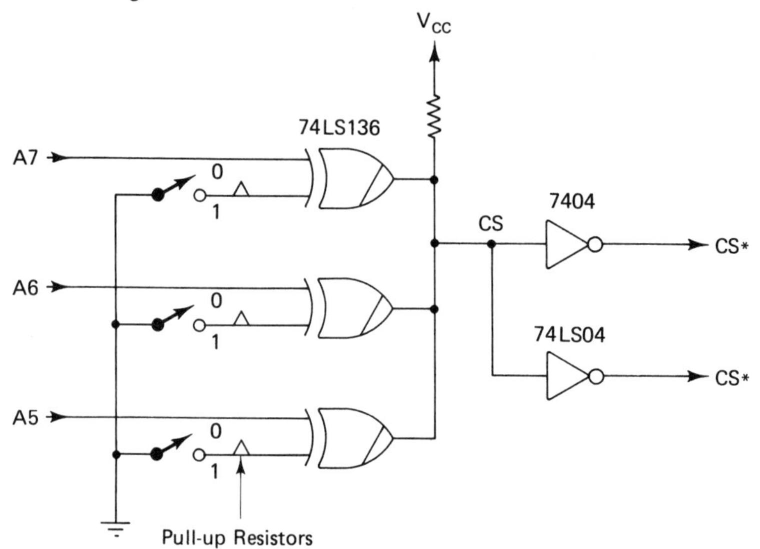

(a) Examine Figure 3.9 to get the following 74LS136 data-sheet information:

I_{OH} = 100 μA max leakage into open-collector when V_{OH} = 5.5 V,

V_{OL} = 0.4 V max when I_{OL} = 4 mA,

V_{OL} = 0.5 V if I_{OL} = 8 mA.

Draw the circuit for a logic LOW at CS:

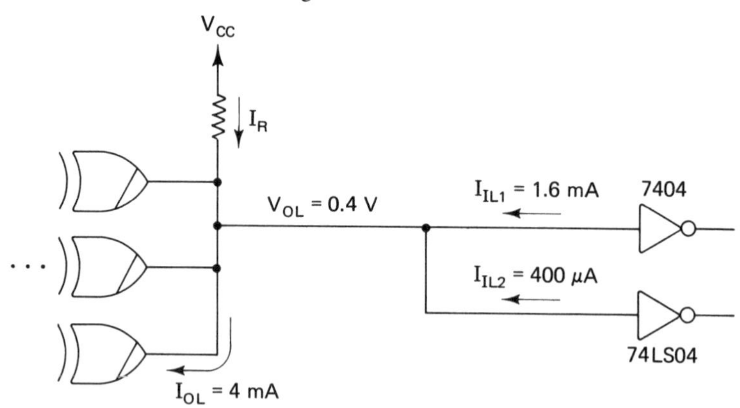

TYPES SN54LS136, SN74LS136
QUADRUPLE 2-INPUT EXCLUSIVE-OR GATES
WITH OPEN-COLLECTOR OUTPUTS

absolute maximum ratings over operating free-air temperature range (unless otherwise noted)

Supply voltage, V_{CC} (see Note 1) . . . . . 7 V
Input voltage . . . . . 7 V
Operating free-air temperature range: SN54LS136 . . . . . −55°C to 125°C
SN74LS136 . . . . . 0°C to 70°C
Storage temperature range . . . . . −65°C to 150°C

NOTE 1: Voltage values are with respect to network ground terminal.

recommended operating conditions

	SN54LS136			SN74LS136			UNIT
	MIN	NOM	MAX	MIN	NOM	MAX	
Supply voltage, V_{CC}	4.5	5	5.5	4.75	5	5.25	V
High-level output voltage, V_{OH}			5.5			5.5	V
Low-level output current, I_{OL}			4			8	mA
Operating free-air temperature, T_A	−55		125	0		70	°C

electrical characteristics over recommended operating free-air temperature range (unless otherwise noted)

PARAMETER		TEST CONDITIONS†		SN54LS136			SN74LS136			UNIT
				MIN	TYP‡	MAX	MIN	TYP‡	MAX	
V_{IH}	High-level input voltage			2			2			V
V_{IL}	Low-level input voltage					0.7			0.8	V
V_{IK}	Input clamp voltage	V_{CC} = MIN,	I_I = −18 mA			−1.5			−1.5	V
I_{OH}	High-level output current	V_{CC} = MIN, V_{IH} = 2 V, V_{IL} = V_{IL} max,	V_{OH} = 5.5 V			100			100	μA
V_{OL}	Low-level output voltage	V_{CC} = MIN, V_{IH} = 2 V, V_{IL} = V_{IL} max	I_{OL} = 4 mA		0.25	0.4		0.25	0.4	V
			I_{OL} = 8 mA					0.35	0.5	V
I_I	Input current at maximum input voltage	V_{CC} = MAX,	V_I = 7 V			0.2			0.2	mA
I_{IH}	High-level input current	V_{CC} = MAX,	V_I = 2.7 V			40			40	μA
I_{IL}	Low-level input current	V_{CC} = MAX,	V_I = 0.4 V			−0.8			−0.8	mA
I_{CC}	Supply current	V_{CC} = MAX,	See Note 2		6.1	10		6.1	10	mA

†For conditions shown as MIN or MAX, use the appropriate value specified under recommended operating conditions for the applicable type.
‡All typical values are at V_{CC} = 5 V, T_A = 25°C.
NOTE 2: I_{CC} is measured with one input of each gate at 4.5 V, the other inputs grounded, and the outputs open.

switching characteristics, V_{CC} = 5 V, T_A = 25°C

PARAMETER¶	FROM (INPUT)	TEST CONDITIONS		MIN	TYP	MAX	UNIT
t_{PLH}	A or B	Other input low	C_L = 15 pF, R_L = 2 kΩ, (See Note 3)		18	30	ns
t_{PHL}					18	30	
t_{PLH}	A or B	Other input high			18	30	ns
t_{PHL}					18	30	

¶t_{PLH} = propagation delay time, low-to-high-level output
t_{PHL} = propagation delay time, high-to-low-level output
NOTE 3: See General Information Section for load circuits and voltage waveforms.

POST OFFICE BOX 225012 • DALLAS, TEXAS 75265

Figure 3.9 (Courtesy Texas Instruments, Inc.)

For the moment, assume that just one 74LS136 is on: any current flowing in the circuit will not be shared by any of the other 74LS136 gates. The current through the resistor is

$$I_R = I_{OL} - I_{IL1} - I_{IL2}$$
$$= 4 \text{ mA} - 1.6 \text{ mA} - 400 \text{ µA}$$
$$= 2 \text{ mA}$$

When the 74LS136 sinks 4 mA, its output voltage rises to a maximum of $V_{OL} = 0.4$ V. Thus, the voltage across the resistor is $V_{CC} - V_{OL}$ or 4.6 V, so the resistor value is

$$R_{\min} = 4.6 \text{ V}/2 \text{ mA}$$
$$= 2300\ \Omega$$

If you allow the 74LS136 to sink 8 mA (resulting in $V_{OL} = 0.5$ V), then you find

$$R_{\min} = 4.5 \text{ V}/6 \text{ mA}$$
$$= 750\ \Omega$$

If you assume that two or three of the 74LS136s might normally be LOW and sharing the 8 mA, then the V_{OL} will be below 0.4 V. This might not be a valid assumption, so it should be verified.

(b) Draw the circuit for a logic HIGH at CS:

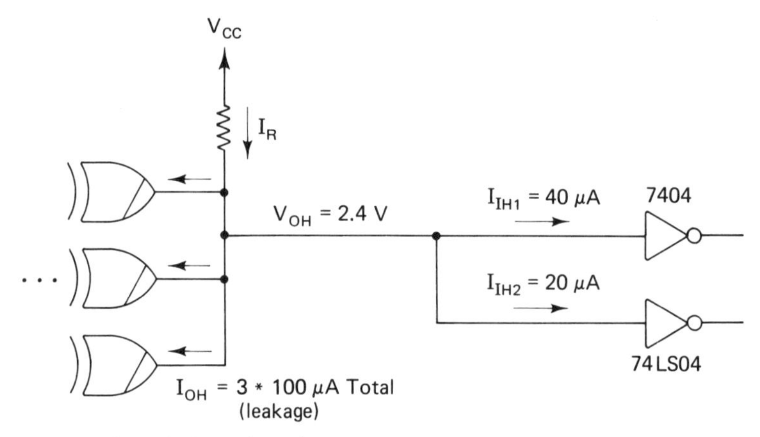

The current through the resistor is

$$I_R = I_{OH}\ (\text{total}) + I_{IH1} + I_{IH2}$$
$$= 300 \text{ µA} + 40 \text{ µA} + 20 \text{ µA}$$
$$= 360 \text{ µA}$$

and the voltage across the resistor is

$$V_R = V_{CC} - V_{OH}$$
$$= 5 - 2.4$$
$$= 2.6 \text{ V}$$

so the resistor value is

$$R_{max} = 2.6 \text{ V} / 360 \ \mu\text{A}$$
$$= 7220 \ \Omega$$

(c) Select some resistor vaue between the minimum and maximum based on speed and power. You might select 2.2 KΩ simply because it works and you have a large stock of them to use.

At some point in the design and development cycle, you should verify your circuit performance with oscilloscope measurements. For the present, you at least have a design within required logic-level specifications.

Rules on Loading and Pull-Ups

1. Use worst-case specifications over operating temperature.
2. Measure actual signal and current levels and compare with design values.
3. Connect unused inputs to V_{CC} through a 1 KΩ resistor. LS-TTL devices, which have internal diodes, can be connected directly to V_{CC}.
4. Examine the loading capabilities of all devices and provide buffering as required.
5. Connect a maximum of ten 74xx inputs to any single 74xx output.
6. Connect a maximum of ten 74LSxx inputs to any single 74LSxx output.
7. Connect a maximum of 20 74LSxx inputs to any single 74xx output. However, double-check the I_{IL} and I_{IH} 74LSxx inputs because they do not all present the same load to a source.
8. Connect a maximum of three 74xx inputs to any single 74LSxx output.
9. Consider open-collector devices if you need wired-AND and OR functions, external relay capability, or bus-interface capability.
10. When you use open-collector devices, calculate a minimum and a maximum pull-up resistance. Then make a selection based on speed and power constraints.

Timing. You find that there is substantially more information in the TTL data sheets than just signal levels and loading: you also find details on signal propagation. The finite time that signals take as they go from one part of your circuit to another can have a profound impact on the performance of your design. The delays in your circuit set a maximum speed for reliable performance, and you need to know how to calculate what that speed is under worst-case conditions.

Propagation Delay. The propagation delay is the time a signal takes to go through a gate or array of gates. If you consider a simple 7404 inverter and drive it with a single pulse, you have a time delay for the output to go high and another delay for the output to go low:

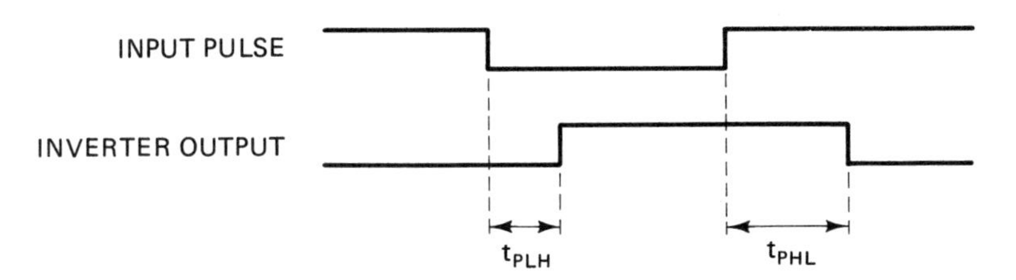

Examine Figures 3.4 and 3.7 to find the propagation times for the 7404, 74LS04, 7405, and 74LS05. Typical and worst-case times are given for both low-to-high and high-to-low output transitions.

You can summarize this data sheet information for the totem-pole and open-collector inverters:

	t_{plh} (ns)		t_{phl} (ns)	
	Typical	*Worst-case*	*Typical*	*Worst-case*
7404	12	22	8	15
74LS04	9	15	10	15
7405 (open col)	40	55	8	15
74LS05 (open-col)	17	32	15	28

In general, if you want to get an estimate of the time it takes a signal to go through your circuit, average t_{plh} and t_{phl} for each device. For example, the average 7404 typical propagation time is 10 ns. For several such gates in series, add the times for each. If you need a worst-case estimate, use the higher worst-case value of t_{plh} or t_{phl}. Thus, the 7404 worst-case propagation time is 22 ns; the worst case for two in series is 44 ns.

The propagation times given in the charts are based on specific test circuits. Compare your circuit to the test circuit to see whether you might expect similar propagation times. For example, the 74LS05 times are taken using the test circuit shown in Figure 3.10. Notice that the t_{phl} for the open-collector 7405/LS05 devices are similar to the

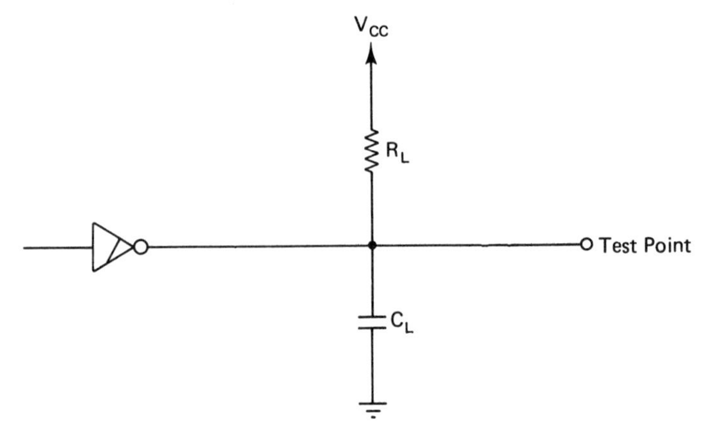

Figure 3.10 Load circuit for open-collector outputs. (Courtesy Texas Instruments, Inc.)

totem-pole gates. This is because the output transistor is quite effective in discharging the load capacitor in a short time. However, the t_{plh} depends primarily on how fast R_L can provide enough charging current for C_L to get the output voltage HIGH. If you need to speed up the propagation time of open-collector devices, you can consider making the pull-up resistor a lower value. In doing that, however, watch not to exceed the maximum I_{OL} for the device; if you go too far, then the output voltage will become higher than V_{OL} and will cause errors in your logic.

The propagation delay through a complete system is an important design consideration. Suppose you have a signal being propagated through a number of gates and need to know how long it takes to get to its destination. Add up each of the worst-case delays, as in this example problem:

Example 5

Look at the address decoder in Example 4 and estimate how long it will take to establish CS* after receiving a valid address.

Solution. See Figure 3.9 for the 74LS136 data-sheet information: $t_{plh} = t_{phl} = 18$ ns typical, 30 ns maximum when R = 2000 Ω and C = 15 pF load.
With the 2.2 KΩ pull-up resistor and two gates loading the 74LS136, you expect that the 30 ns worst-case time is reasonable. Add the 22 ns worst-case time through the 7404 inverter, and you conclude that CS* is valid about 52 ns after a valid address. Typically, however, you expect a propagation time of 18 + 10, or 28 ns.

Clock Timing. In a large digital system, there are many sequential devices interconnected that need synchronization using the same clock signal. If your clock has a single 7404 or other 74xx output device, it cannot drive more than 10 74xx loads; however, there are several ways to boost the drive capability of your clock. The first way, as shown in Figure 3.11(a), is to parallel the gate inputs and outputs of a single package; avoid using separate packages because the different switching times will cause problems with noise transients. The second way, Figure 3.11(b), is to use a clock driver such as the 7437 to provide a drive for up to 30 74xx loads. For still more drive capability, multiple sections of a 74S37 can be used as in Figure 3.11(c).

When you divide the clock load between two or more drivers as in Figure 3.11(c), then you cause a potential clock skew problem in your system. Clock skew is the difference in time between two clock signals. Suppose your system clock goes to two 7404 inverters, each of which goes to two different parts of your system. As shown in Figure 3.12, one 7404 has 3 loads, and the other has a full 10 loads. Both 7404s are different and have propagation times that can be from the worst-case 22 ns to better than the typical 10 ns. Even if the gates are identical, the difference in loading affects the propagation times. Consequently, you might find a clock skew of 12 ns or more between the two clock circuits.

You might improve the circuit by changing from the 7404 to the Schottky 74S04, which has a worst-case propagation time of 5 ns. The resulting clock skew would be less than 5 ns, typically less than 3 ns. The tradeoff you make in going to a Schottky device is

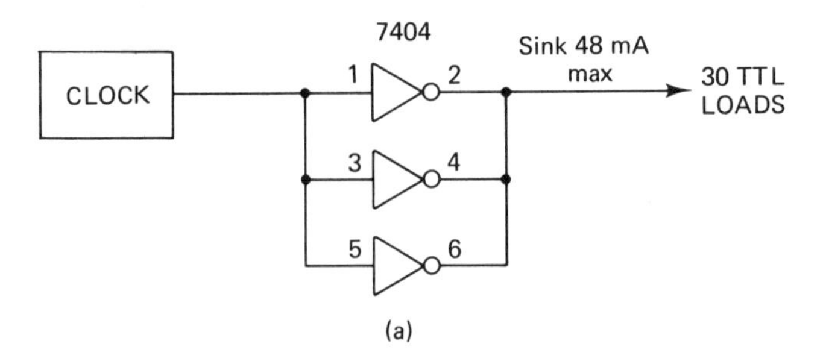

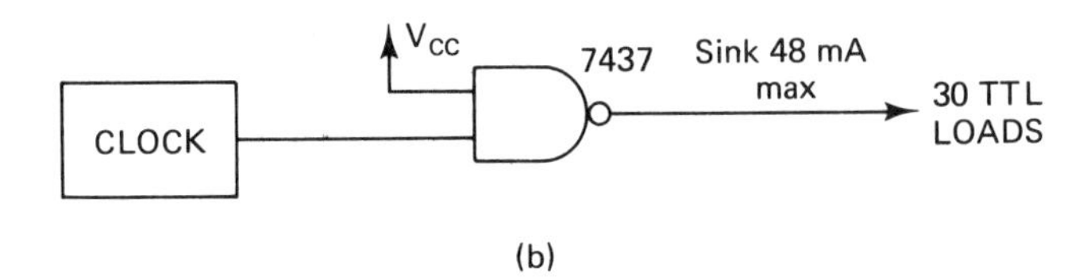

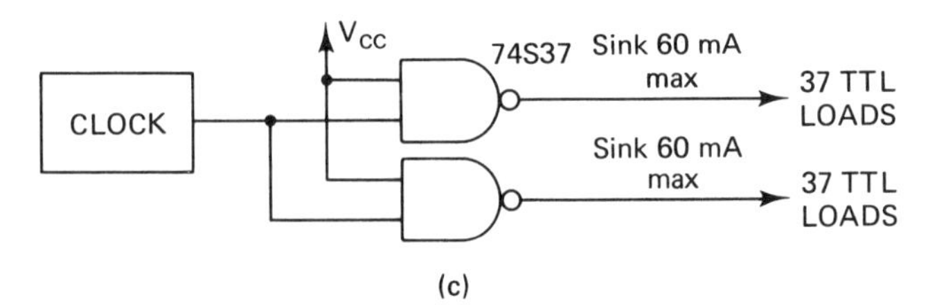

Figure 3.11 Several clock driver circuits.

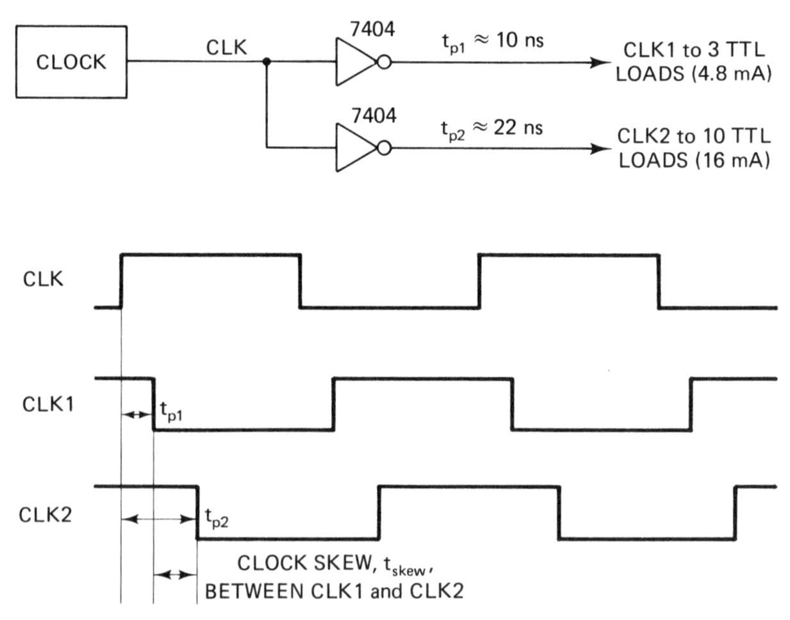

Figure 3.12 Clock with two clock drivers that cause a skew between CLK1 and CLK2.

an approximate doubling of the supply current to power the IC. If you have a fairly simple system, you probably would not notice the skew at all and have no need to resort to a 74S04. If you use a Schottky device, watch that you do not cause problems with noise: the very fast switching speeds can not only affect the 5 V supply but can also couple energy into adjacent circuits.

Sometimes a complementary clock is needed in the system, as depicted in Figure 3.13. Using a simple 7404 inverter to obtain an inverted clock could cause a clock skew of up to 22 ns. However, the 74265 quad complementary-output circuit provides the two clocks with a typical skew of 0.5 ns and a maximum skew of 3 ns.

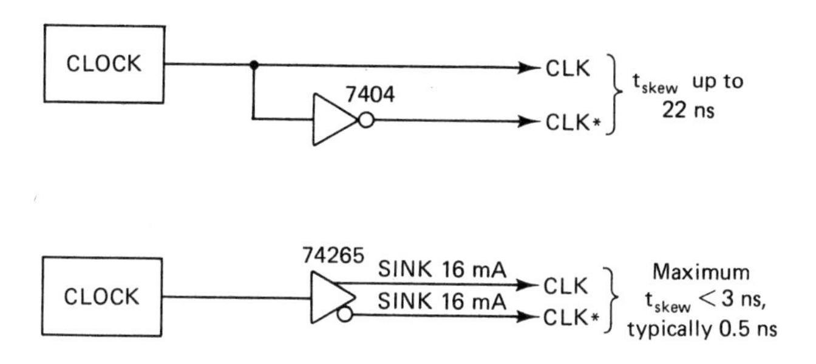

Figure 3.13 Rather than using a 7404 to get a complementary clock signal, use a device designed for a minimum output skew.

Setup and Hold Times. Each pulse of the system clock signals the beginning of a number of events in a digital circuit. It takes a finite time for each event to propagate through the system to get all the devices ready for the next clock pulse. That is, the proper data must be present for a certain time before the clock pulse arrives. You can define this as the "setup time."

> *Setup Time* is the time interval preceding the clock pulse that a clocked device must have valid input data. A negative setup time means that valid input data may be applied after the clock pulse and still be properly recognized.

Likewise, it takes some time after the clock pulse arrives for the input data to be acted upon by a sequential device. This is referred to as the "hold time."

> *Hold time* is the time interval after the clock pulse that a clocked device must have valid input data. A negative hold time means that the input data may be removed prior to the clock pulse and still be properly recognized.

There are also requirements on the duration of the clock pulse itself. Normally, most of the flip-flops you will use in your TTL designs will be edge-triggered. That is, the clock transition from LOW to HIGH (positive edge) or from HIGH to LOW (negative edge) determines the exact moment to change state. Although the clock edge triggers the flip-flop, the clock pulse must remain high or low for a certain time. Consequently, the required pulse width is specified in the device data sheet.

Pulse width is the time between the leading and trailing edges of a pulse. A width may be specified for either or both of the high and low portions of the clock cycle.

Example 6

What are the requirements for setup and hold times of the 7474 D flip-flop? What is its maximum clock speed? How does this relate to the propagation time?

Solution. When you examine the data for the positive-edge-triggered 7474, you find:

t_{su}	20 ns	setup time
t_h	5 ns	hold time
t_w	30 ns	pulse width, clock high
	37 ns	pulse width, clock low
t_{plh}	14 ns	typical propagation time
	25 ns	worst case
t_{phl}	20 ns	typical propagation time
	40 ns	worst case

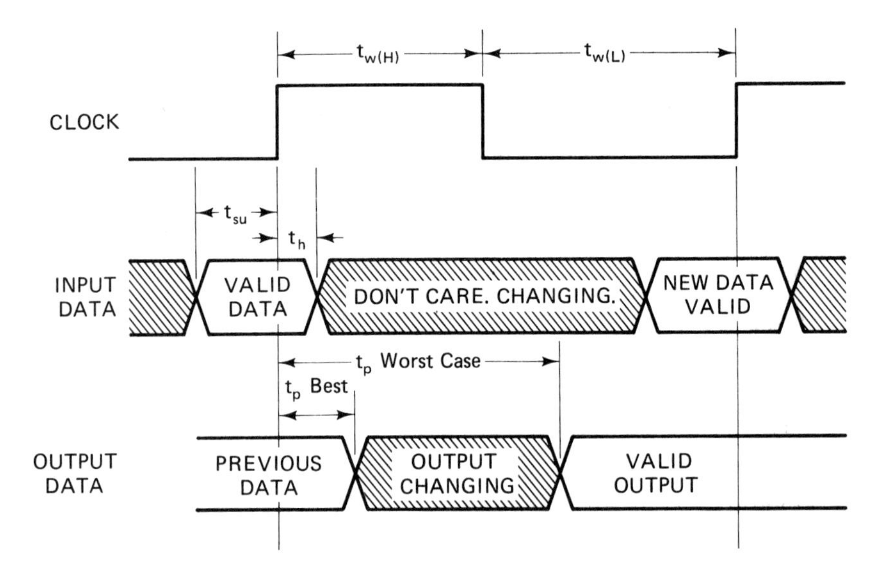

If you add the clock pulse-width times specified (30 and 37 ns), you can calculate the 7474 maximum clocking frequency. In this case, the times add to 67 ns, which corresponds to 15 MHz.

Example 7

Clock-speed calculation: How fast can the clock run and still drive this circuit reliably?

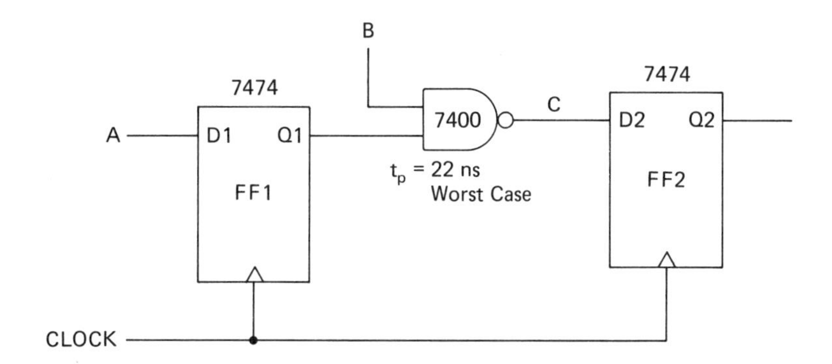

Solution. Assuming that input A setup and hold times are met and that input B is not changing, you can add the worst-case times through each of the devices and draw the timing diagram.

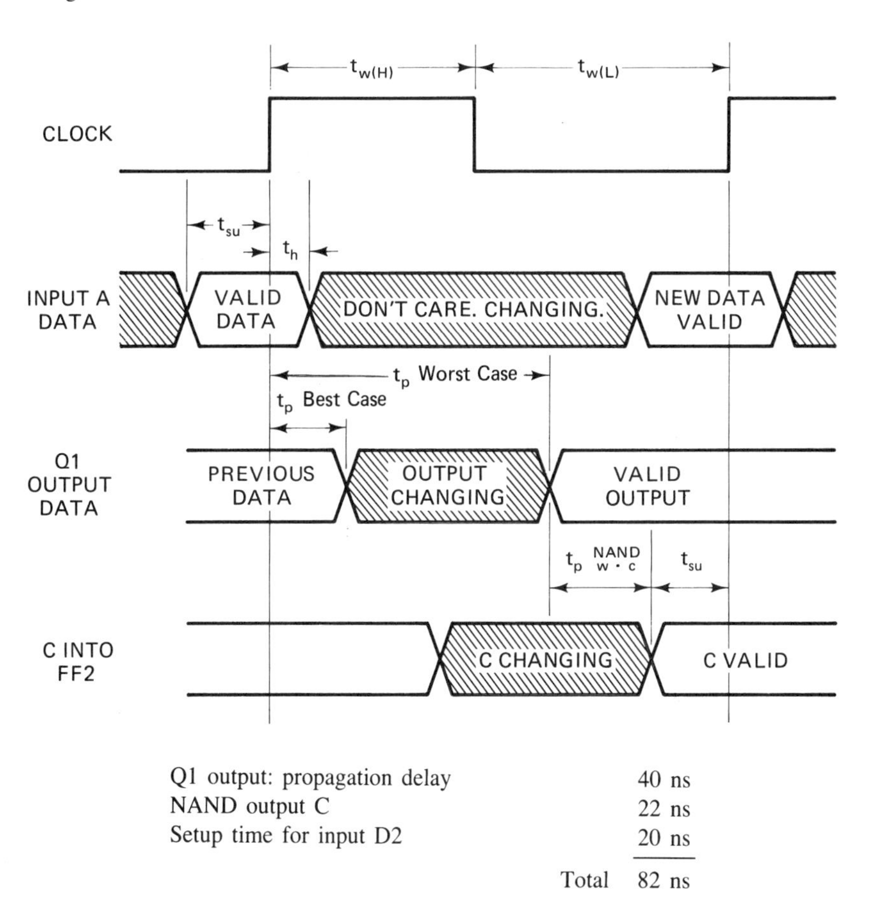

Q1 output: propagation delay	40 ns
NAND output C	22 ns
Setup time for input D2	20 ns
Total	82 ns

Therefore, the maximum clock speed should be less than 1/82 ns or about 12 MHz. Suppose that there are five NAND gates between the flip-flops instead of one. The 22 ns above would be 5 × 22 or 110 ns for a total of 170 ns. The maximum clock would then be just under 6 MHz. Anything faster might not operate reliably. In general, you can calculate the maximum clock speed using the worst-case propagation through the *longest* signal path.

Example 8

Example clock-skew calculation: What is the maximum tolerable clock skew for the following circuit?

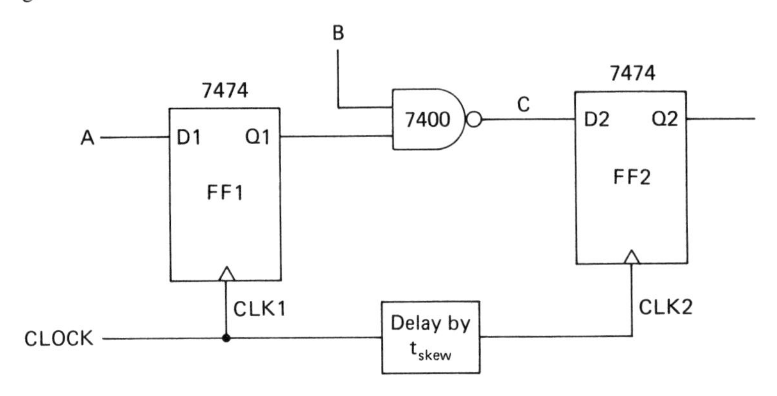

Solution. Draw the timing diagram as before to obtain the arrival time of the data at output C of the 7400. When you calculate the maximum skew, however, use the *fastest path* between the two flip-flops; *the worst-case propagation time for the NAND is actually the favorable case.* You can estimate the fastest time as the value given in the data sheet as typical; or to be conservative, you might estimate a still faster time for the NAND.

If the maximum clock-skew time is exceeded, then you will be clocking the second flip-flop when the NAND output C could be in a state of change. The maximum clock skew is the best propagation time through the first flip-flop plus the best time through the NAND minus the hold time of the second flip-flop.

$$\begin{aligned}\text{Max } CLK2 \text{ skew} &= t_p \text{ (best thru } FF1) + t_p \text{ (best thru NAND)} - t_{\text{hold}} \text{ (max } FF2) \\ &\approx 14 \text{ typ} + 7 \text{ typ} - 5 \text{ typ} \\ &\approx 16 \text{ ns or less}\end{aligned}$$

Rules on Clocks and Timing

1. Estimate propagation time by taking the total of all the worst-case times in the data path.
2. Estimate typical propagation times by taking the average of t_{phl} and t_{plh}.
3. Adjust estimated times depending on how heavy the load capacitance is when compared to the load used to obtain data-sheet times.

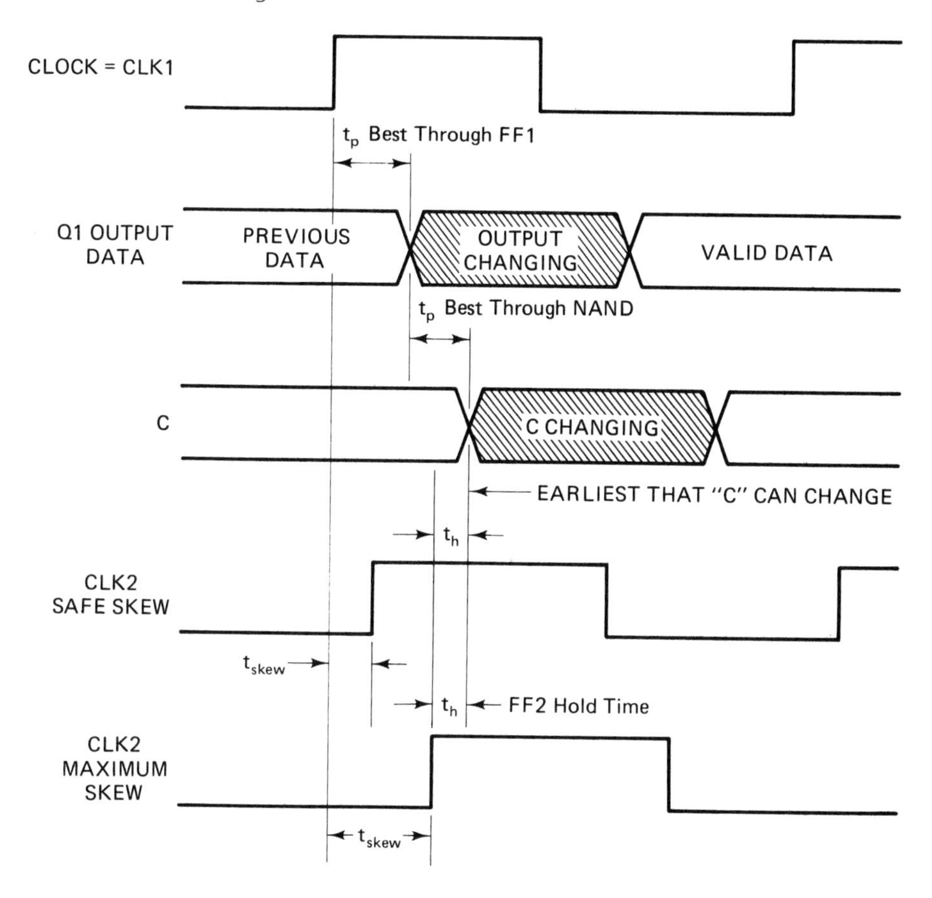

4. Calculate the maximum clock speed using the longest signal path between two flip-flops (or similar clocked devices). Use the worst-case propagation times for logic gates along the path.
5. Calculate the maximum allowable clock skew using the shortest path between two flip-flops (or other clocked devices). Use the fastest propagation times for any logic gates along the path.
6. Verify that clock speed is not excessive.
7. Verify clock skew through the system.
8. Use pull-up resistors on any unused logic gate inputs to avoid degradation in package propagation time. Doing this will ensure best switching times and minimum noise susceptibility.
9. Verify that clock high and low times are met. This might be part of the duty-cycle specification on a DIP clock you buy as a complete unit.
10. Verify that setup and hold times are being met.
11. Review the timing diagrams related to all LSI devices.

12. Use processor wait-states if necessary to slow the system down for certain operations rather than run the risk of sometimes having a system failure.
13. Select or design a clock for a fast rise time to minimize skew.
14. Shield the clock line or keep it physically isolated or at right angles from other logic paths if possible.
15. All devices should respond to the same clock. Tie all clock inputs together.
16. Do not use an inverter to get a complementary clock signal.
17. Do not use unclocked sequential devices.
18. Do not operate logic at its maximum speed.
19. Find the critical timing path that has the longest propagation delay and verify that system timing is proper.

Noise. All the TTL devices you use in your designs are susceptible to possible malfunction because of noise. This noise is a corruption of the logic signal levels so that, for example, a signal with a logic HIGH might be misunderstood as a logic LOW.

The noise margin is the maximum voltage disturbance that can be added to the input of a gate; any disturbance greater than the margin might cause an unwanted change in output. When you first examined the TTL allowable input and output voltage ranges in Figure 3.3, you noticed that the TTL outputs are specified less than 0.4 V and greater than 2.4 V. The inputs, however, accept less than 0.8 V or greater than 2.0 V. This difference is referred to as the "noise margin." For the TTL logic levels, the noise margin is 0.4 V.

You can see how noise affects the signal, as shown in Figure 3.14. This noise is superimposed on the desired signal and could cause an unwanted change in the output. Notice that the noise is high-frequency spikes, or glitches. In (a), a brief burst of noise exceeding 0.8 V caused a momentary transition in the output (b); likewise, an instantaneous drop below 2.0 V in (a) caused a brief output transition in (b). To respond to such a sudden noise impulse, the logic device must have been quite fast (say a 74S04). In general, if your logic family is slow, then it will not be prone to respond to high-frequency noise. It will, however, still respond to excessive lower-frequency noise.

Noise in the digital system can come from many sources. One of the primary sources of TTL noise comes through the +5 V supply lines; this noise is caused by various gates switching from one logic level to another. When you first examined totem-pole devices, you saw the two output transistors as an advantage to boost the device speed: turning on one or the other could quickly change the state of the output. In reality, the two transistors are both on for a brief instant when they switch, and this causes a heavy current surge from the power supply. Unless the power supply and wiring to the TTL device is substantial, the supply voltage V_{cc} at the device will drop. This causes drops at nearby ICs and might cause misinterpretation of data levels. To help mitigate the switching noise, decoupling capacitors are used near every several IC packages. Typically, a 0.01 μF capacitor is used for two ICs; this capacitor value is not critical, nor is it critical to use a capacitor for every two devices. Some designs use decoupling capacitors between V_{cc} and ground built into the IC sockets themselves.

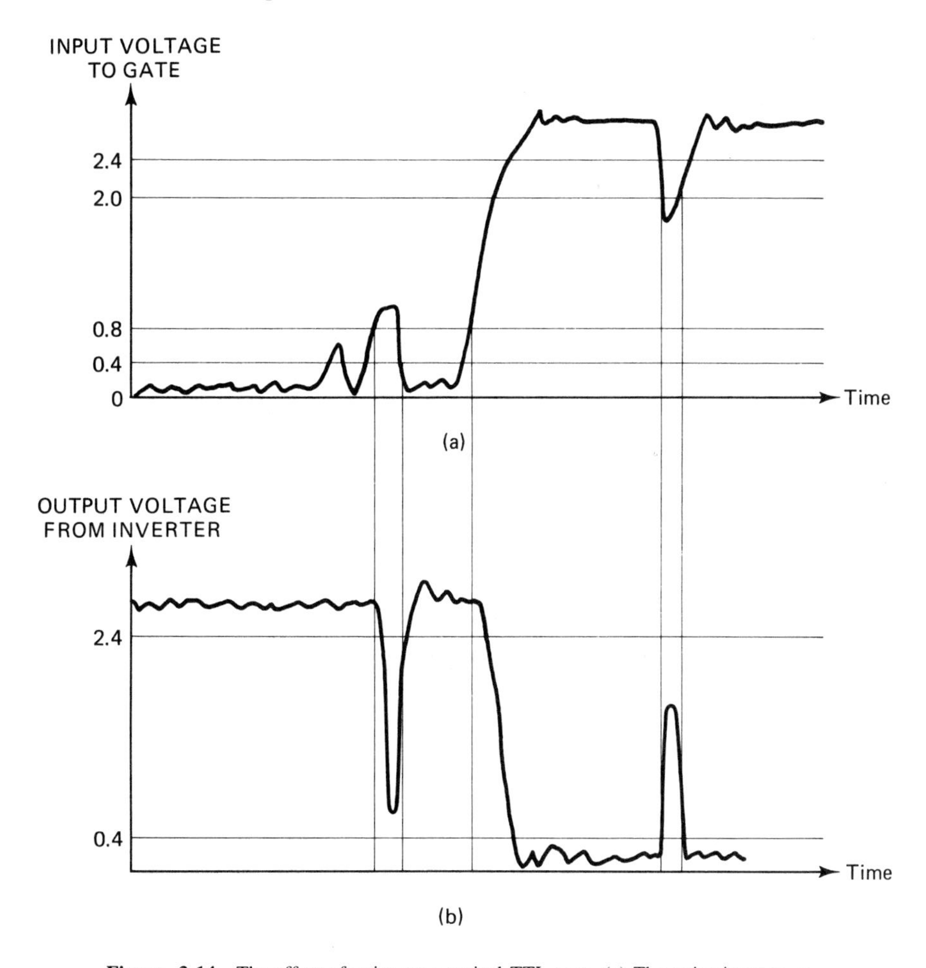

Figure 3.14 The effect of noise on a typical TTL gate. (a) The noisy input to an inverter with two noise bursts exceeding the noise margin. (b) The possible output of the same inverter. Drawing assumes zero propagation delay and that the gate switches immediately on entering the transition region.

Another source of noise is the clock itself. One desirable feature of the clock is that it have a fast risetime to cut down on clock skew problems. However, this fast transition can be easily radiated and cause crosstalk, or extraneous signals in nearby circuits. Shielding the clock lines or using twisted-pair cable can help, but on a circuit board a good layout is most effective in minimizing the crosstalk. Rather than run the clock near data lines, keep it separate; if possible, run the clock on the opposite side of the board at right angles to any data lines.

Rules on Noise

1. Tie unused gate inputs to V_{cc} or use pull-up resistors.
2. Assert communication lines to the system low-true for best noise immunity.
3. Use 0.01 μF decoupling capacitors every two or three ICs.
4. Design with the slowest acceptable logic family.
5. Keep clock lines away from data lines.
6. Use a ground plane and heavy ground and power leads.
7. Keep connections to all parts as short as possible.
8. Use wide, low-impedance conductors on PC boards.
9. Avoid crosstalk with twisted-pair or shielded cable.
10. If using double-sided PC board, run circuits perpendicular to each other on the top and bottom. Use ground and power planes if using multilayer boards.
11. Shield the clock line or keep it isolated from other circuits.
12. Verify that the noise level is less than the design margin.

General Design Rules

1. Design for testability: include push-to-test or self-test capability.
2. Do modular system design so circuits can be developed and debugged by stages.
3. Design and test the power supply module first.
4. Design utility modules such as bus buffers and address decoders early.
5. Use mixed logic symbols in all schematics for quick comprehension.
6. Examine all LSI devices to see if buffering should be provided.
7. Watch for possible bus contention problems.
8. Use worst-case specifications.
9. Use proven circuits; save your time for creative problem solving.
10. Verify that the system works over a reasonable temperature range even if there is no explicit temperature specification.

3.1.2 Software Design Rules

Regardless of the language or processor you use, your programs can be treated the same as hardware; that is, programs can be interconnected and used as building blocks to solve a particular problem. Visualize a program as you would an LSI device: it has inputs, outputs, and performs a particular function in the system. In the same sense, your programs can be modularized and used to save substantial design time as your project evolves.

In order to make your programs into modules that you can call as needed, you should do a top-down design of your software just as you did your hardware design. This design can be easily drawn in the form of a structure chart, as shown in Figure 3.15.

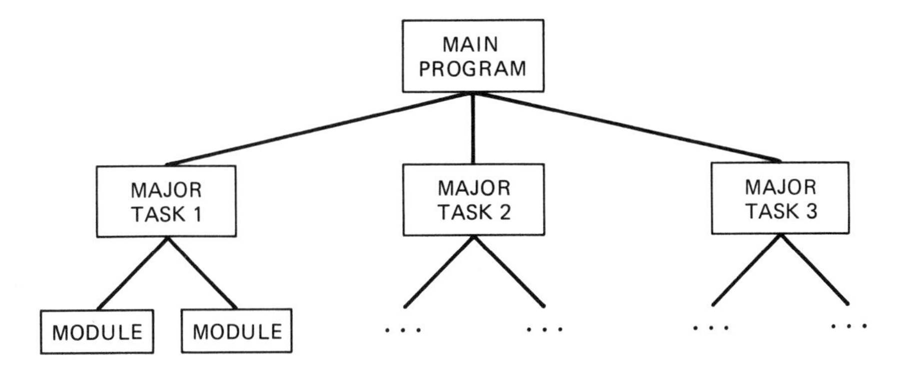

Figure 3.15 The form of the structure chart used in a top-down program design.

Consider each block as a functional module within the system and define the purpose of each. Start with the purpose of the main program: what does it do and what are its inputs and outputs? Then, for the main program to accomplish its purpose, a number of supporting tasks are necessary. The process becomes more detailed as you get down to the next level in the structure chart.

The easiest way to visualize the decomposition process into modules is to think of the main program as in the form of a menu of tasks that can be selected. Each of the tasks is a subroutine within the program and can be selected from the menu for execution. Once a task has been selected, it needs to call on a number of supporting modules or routines during execution.

If you think of each of the blocks of the structure chart as subroutines, you can see the advantage of making each module as straightforward and specific as possible. With one function per module, you can combine them in just the same fashion as you design hardware. Not only that, but you can write the code for the module and thoroughly test it before "building" it into the larger body of the main program.

Rules on software design

A module should:

1. Have one purpose.
2. Be as general as possible for use in other applications.
3. Be reliable. (Provide worst-case test data to demonstrate.)
4. Have a length less than two pages.
5. Contain one statement maximum per line of program.
6. Have one entry point and one exit point.
7. Be internally documented with ample comments.
8. Contain a heading with: name of module, date, revision number; list of external

modules and variables; public labels and variables; description of input and output variables; and note of any registers that are changed during a call to the module.

In general:

1. All constants should be EQUated at the beginning of the main program or in a library related to the appropriate module.
2. Routines, parameters, labels, storage area, and all calling procedures should be consistent with each other.
3. Avoid absolute addresses if possible. Use relative addressing. Program for easy relocation of module. Ideally, your module should execute wherever you place it in memory.

3.2 DESIGN HEURISTICS

Your design up to this point has probably been somewhat conventional: define the problem and specifications, synthesize several possible solutions to the problem, select the best solution, and then implement it. Although these are the general steps in the engineering design process, they seem somewhat pedestrian and lack vigor. Where is the creativity and flexibility to innovate?

In reality, when you first begin a design task, you do not understand the problem. You cannot decide which specifications are critical to the design or even decide which job to do first. Why not follow some rule of thumb, or heuristic, that can help you get moving in the right direction as you feel your way into a better understanding of the problem?

By its very nature, a heuristic is a guide based on past experience and common sense. A heuristic is not a rule that can be conclusively proven, because the heuristic might not even be correct in certain situations. Therefore, when you use heuristics to help speed your design work, remember that they do not guarantee a correct answer and that you must still work out a proper design. The advantage of the heuristic is in helping you decide on which design might be worth your time and effort.

The following list of heuristics are a collection of useful rules for general engineering as well as digital design. Some are obvious, some are obscure, but many should be useful.

3.2.1 Engineering Heuristics

1. Don't reinvent the wheel: read data sheets and application notes.
2. Reduce your problem to something you've already solved before.
3. If you can't meet the specs, negotiate; don't hide the problem.
4. Always have an answer; you have to start somewhere.
5. Change one variable at a time when you adjust your design.
6. Develop circuits and programs module by module; debug as you go.
7. Build a quick simple circuit for experimentation; understand it.

8. Keep your designs simple.
9. Use LSI devices whenever possible.
10. Talking aloud to yourself helps spot errors.
11. If you find you made a mistake, figure out why.
12. Solve the right problem.
13. Act rather than react: don't spend your day fighting fires.
14. Read the fine print at the bottom of data sheets.
15. When in doubt, don't guess; look it up and be sure.
16. On time management:
    Keep a daily do-it list with priorities for each task.
    Do critical or difficult tasks as soon as possible.
    Schedule unfinished tasks for a definite day in the future.
    Keep a time log of your work and review your progress.
    Don't procrastinate—a project gets late one day at a time.

3.3 SUMMARY

Design rules are the procedures or conventions established to direct your project. These design guidelines, rather than being confining, can help make your design effort more orderly and technically sound. They also help make your work consistent with the designs of other team members working on the same project. This is important so that you can complete your project within your available time and budget.

One of the nice features of digital design is that the circuits can be easily connected in building-block fashion. When you work out a logic function, for example, you can usually implement it by simply interconnecting the required gates. When you do this, however, you must be careful to stay within the TTL signal level and loading specifications. You must consider how many totem-pole devices you can connect before the signals become too degraded; likewise, you must calculate the size pull-up resistor for open-collector devices so that its output is satisfactory.

Timing is one of the most important issues in designing your digital system. When you include propagation time in your design, you find that it can affect a number of different parts of your circuit. This becomes especially critical when you select a clock frequency that pushes the system to its speed limit. Because not all components perform the same, one circuit might work at a high speed while another might not. For all your designs, therefore, be sure to design for the worst-case component tolerances.

Noise can cause a number of strange malfunctions at odd times during circuit operation. The noise can come from external sources or from inside the digital system itself. The best design approach to take is to design so that you have a noise margin in your logic levels. For example, design your gates for an output LOW of less than 0.4 V; that way, any inputs connected to it will not mistake it for a logic HIGH unless the output exceeds a total of 0.8 V.

Software can be designed following the same general approach that you use when you design hardware. You can do a top-down modular program design just the same as you would design using LSI devices in a hardware system. Each of the modules has an input, an output, and performs a special function. A structure chart is an easy way to visualize how the various software modules in the system relate to one another. As you write each of the modules, you can test and debug them as you go, just the same as you test and debug hardware.

Engineering heuristics are rules of thumb or guides that you can use in a somewhat intuitive sense to solve engineering problems. They cannot be proven, and some might not even work in any particular situation. They can, however, speed your design work by giving you a feel for what might be important in the design. At the very least, heuristics can usually get your initial calculations approximately correct. Instead of getting bogged down in a deep theoretical design issue, you can speed your work with the common-sense reality of the heuristic.

By applying both technical design rules and engineering heuristics, you can work much more effectively to finish a design problem. The guidelines will help you do a technically sound design that is consistent with the design work of other team members. They will help make the work go faster and, as you use more heuristics, make design engineering an enjoyable challenge to your creativity.

EXERCISES

1. What is the maximum operating temperature for a 7404? Express it in Fahrenheit as well as centrigrade.
2. What is the maximum worst-case supply current required by a 7404 package? Compare this to the worst-case 74LS04.
3. Suppose the output of a 74LS00 is accidentally shorted to ground, and your circuit causes the device to try switching to a HIGH output.
   a. How much current will flow maximum?
   b. Which way?
   c. How long?
4. Consider the circuit in Example 3. Suppose you built the circuit and accidentally forgot to hook up the pull-up resistor. Will the circuit function? How well or poorly? Build up a quick test circuit to see what happens. Explain your answers. Hint: This problem is *not* as trivial as it might appear.
5. Ten LS gates are connected to the output of a single 74LS04, but you need to add two more LS gates because of a circuit change.
   a. Predict the effect on the logic HIGH and logic LOW at the output of the 74LS04.
   b. Can you use the circuit as is, with 12 gates connected? What compromises are you making?
   c. Suppose you decide not to connect 12 gates to the 74LS04. What alternatives are open to you? Sketch a circuit and defend its technical merit. State your assumptions and justify them.

6. You want to connect a 7407 output to the RESET* input of a 68000 microprocessor.
   a. Calculate a minimum and maximum value for a pull-up resistor. Assume the 68000 input looks like a 74LS04.
   b. Assume three LS gates are also connected to RESET*. Calculate a minimum and maximum pull-up resistor. Make a choice.
   c. Why not use a totem-pole device instead of the open-collector 7407?
   d. Suppose the 68000 executes the RESET instruction and the RESET* pin is suddenly pulled LOW (V_{OL} = 0.5V when I_{OL} = 5mA). Calculate 6b once again.
7. Consider text Example 3 of the 7405 connected to a 7404: the two resistances were calculated at 320 and 8970Ω. You decide to try saving current by using an 8.2K pull-up.
   a. Calculate the risetime as the output goes from low to high. (Assume that C = 5pF load.)
   b. What is the risetime for a 330Ω pull-up resistor? (Assume that C = 5pF load.)
   c. Your system runs on a 12 MHz clock. Sketch a clock cycle and a "compromise" signal from the 7405. What value resistor will suffice to produce a reasonable signal?
8. Four 7400 gates are in series in a signal path you traced in a digital system.
   a. What is the worst-case propagation delay? Typical?
   b. You see that each gate output only has two 7400 loads. Estimate the worst-case propagation delay.
9. The worst-case delay in Problem 8b is not acceptable. You must have your signal guaranteed valid within 20 ns to assure enough setup time for the following clocked device.
   a. What can you do?
   b. What tradeoffs are you required to make?
10. How fast, worst-case, can you clock a 74LS109A?
11. Sketch the clock, input data, and output data for the 74LS109A by following the pattern in Example 6. Assume a 16 MHz clock and draw approximately to scale.
12. You have a circuit similar to Example 7 except that you have an 74LS109A instead of a 7474. How fast can the clock run?
13. Suppose you modified your design and could run the 74LS109A at 8 MHz in the circuit in Example 8. How much skew can you allow in the clock circuit?
14. Explain how a Schmitt-trigger device such as the 74LS14 could help prevent noise disturbances in a circuit.

FURTHER READING

COMER, DAVID J. *Digital Logic and State Machine Design*. New York: Holt, Rinehart and Winston, 1984. (TK 7868.S9C66)

FLETCHER, WILLIAM I. *An Engineering Approach to Digital Design*. Englewood Cliffs, NJ: Prentice-Hall, 1980. (TK 7868.D5F5)

FORBES, MARK, and BARRY B. BREY. *Digital Electronics*. Indianapolis, IN: Bobbs-Merrill Educational Publishing, 1985.

HAYES, JOHN P. *Digital System Design and Microprocessors*. New York: McGraw-Hill, 1984. (TK 7874.H393)

KLINE, RAYMOND M. *Structured Digital Design*. Englewood Cliffs, NJ: Prentice-Hall, 1983.

PEATMAN, JOHN B. *Digital Hardware Design*. New York: McGraw-Hill, 1980.

The TTL Data Book. Vol 2. Dallas, TX: Texas Instruments, Inc., 1985.

WIATROWSKI, CLAUDE A., and CHARLES H. HOUSE. *Logic Circuits and Microcomputer Systems*. New York: McGraw-Hill, 1980. (TK 7868.L6W5)

WINKEL, DAVID, and FRANKLIN PROSSER. *The Art of Digital Design*. Englewood Cliffs, NJ: Prentice-Hall, 1980. (TK 7888.3W56)

FOUR

Design Documentation
Putting Ideas on Paper

When you begin your project, you have many ideas for solving the various parts of your design problem. At the beginning of your work, you should outline your project goals, objectives, strategy, and plan in a mini-proposal. Is this the end of the paperwork? No, you know that either your manager or your professor expects a report of some kind at the end of the project. How can you be ready to easily prepare a report on your work?

This chapter demonstrates how to record your daily progress systematically and how to document the final results of your project. This documentation is important because someone else such as team members, manager, professor, or customer needs to understand and to use your design or product. You benefit as well: as you record your progress each day, you can review your work and learn from your mistakes. A daily review gives a better direction to your work.

You can record your daily progress most effectively by using a laboratory notebook. The lab book records what design and testing you do each day. In a larger sense, it contains everything about your project from its conception to the final report. There are many ways to keep a lab book, but the most important point to remember is that it is your ''idea book,'' where you jot down your thoughts each day as your work progresses.

You can fully document your final results in the form of a technical design report to an engineering manager, a design report to a customer, or a semester project report. We will use the form of a technical manual because it contains all the information needed to install, operate, understand, and troubleshoot a product.

In order to present your design information clearly and consistently, you should follow the drawing guidelines given in the Appendix when you prepare your technical manual. This chapter will help you explain the drawings you prepare for hardware designs as well as for software designs.

4.1 LABORATORY NOTEBOOK

Your laboratory notebook is your workbook where you record the tasks you do each day. All your ideas and thoughts on your project should be included in it, beginning with the project conception, through the written proposal, and ending with the final report. Consider it your "idea book" where you write down everything that seems even remotely useful.

Physically, your lab book should be a bound notebook with numbered pages; it should not be a loose-leaf notebook because pages can be removed and easily lost. All writing should be in ink rather than in pencil. Date each page as you use it, and if you are involved with some patentable ideas, have a colleague witness and sign the pages with you. This is important because your lab book is evidence if you ever need to prove in court when you first worked on an invention.

Because your lab book is so special in this regard, be careful never to remove pages or obliterate any material. If you make an error, never erase it or blank it out: line it out and place the correct data beside the error. If a page is in error, put a large "X" through the page. Later, as you review your progress, you can see your errors and perhaps avoid making the same kind of mistake in the future.

Your laboratory notebook should be neat. However, never take data on a scrap piece of paper with the idea of later copying it neatly into the lab book. Loose paper gets lost easily; even if not lost, copying into the lab book might cause errors. Your lab book must always contain original data.

When you write in the lab book, do most of your work on the right-hand pages. Save the left-hand pages for graphs, for equations you might want to find easily, or for marginal notes of any kind. As you work on some of the creative parts of design, you might find the left pages useful for rough sketches of your ideas.

You might find it helpful to tape photocopies of manufacturers' data sheets in your lab book so you can easily refer to the information. In addition, you might need to use your lab book in several years when you do a similar design; by that time, the data manual you used originally might be replaced or lost. Your lab book should be as complete a record as possible of all that goes into your design; it should be able to stand alone.

If you find a technical article that is especially relevant to the lab work you are doing, tape it into your lab book for easy reference. Be sure to note its bibliographic information. Generally, however, you will find it more helpful to collect relevant articles and file them by topic rather than putting them in the lab book. Depending on the nature of your project, you might want to keep the lab book and the article files together and not mix the articles in with any others you might be saving.

The organization you follow when you actually enter information in your lab book depends on the nature of the problem. If you have a short "experiment" to perform, you will find the outline shown in Table 4.1 helpful. Notice that it follows the pattern of problem solving presented in Chapter 1. Does it make sense to use Table 4.1 for a large project that might take weeks to complete? Yes, because with a sizable job, it is easy to lose sight of your objectives and to waste time working on some minor detail. As with any design work, the outline is only a guide and should be modified to fit your particular situation.

TABLE 4.1 AN OUTLINE OF A TYPICAL LABORATORY EXPERIMENT.

Problem Statement	Briefly state an overview of what you are trying to accomplish. What is the problem or purpose of the experiment. Significance?
Objectives	List specific measurable outcomes of your experiment.
Background	Outline of relevant theory or practice. Include references for sources of information.
	Relate present problem to past work.
	Do an analysis of the present problem. (Do diagrams, equations, derivations, simulations.)
	What results are expected? (The analysis should give preliminary answers to each of your objectives stated above.)
Experiment Design	Plan how to obtain the data necessary to verify or disprove your analysis.
	Sketch equipment configuration and test circuits you need to attain each objective above.
Experiment	For each objective, connect equipment in accordance with your plan and make measurements and observations. Put your data in tables if appropriate.
	Note the model and serial number of the equipment you actually use for the experiment in case you need to verify data or expand the scope of your experiment later.
Evaluate Results	Analyze the data collected. What does the data indicate? Plot graphs of data if appropriate. Are there interrelations that explain what happened?
	For each objective, interpret the results and compare with the expected results. Why are there any differences between actual and expected?
Conclusions	What is your answer to the original problem?

If you begin your lab book with your original project concept and thoughts leading up to your proposal, then you will be able to focus quite clearly on your objectives. Break up a large project into manageable smaller tasks, work each task as shown in Table 4.1, then write a short summary for each task as you complete it. Each of these summaries can be used later as part of your progress reports or as part of the final project report.

Keeping your lab book neat and complete takes time. However, that time is well worth the effort. In addition to its value as a record of your designs and data, it is your daily idea book of useful information.

4.2 TECHNICAL MANUAL

The final results of your project can be reported in a number of ways depending on the reader and how the report will be used. A report intended for management, for example, will be different from a report directed primarily to engineers. A report in the form of a technical manual for engineers will be adopted for our use here because of its technical completeness. We will not be concerned with trade secrets and whether or not the product design is proprietary; delete sections your company normally withholds.

As shown in Table 4.2, the complete technical manual contains the information nec-

TABLE 4.2 AN OUTLINE OF THE MAJOR SECTIONS OF A TECHNICAL MANUAL.

Title Page	Title of project or product, author, date.
Introduction	Purpose of the product: What problem was solved? Importance: Background information and how the product relates to prior designs. Features: Description of unit and how it solves the problem.
Installation	System Configuration Hardware setup and requirements Equipment interconnections Switch and jumper settings Location drawing of switches and jumpers Function table of switches and jumpers Connectors Location and assignments Table of signals at each connector Software organization and requirements Memory map I/O map System checkout instructions
Operation	Operating instructions Operating Difficulties (How to Resolve)
Circuit Description	Theory of Hardware Operation Block diagram of system and modules Explanation of functional modules Timing diagrams
Software Description	Theory of Software Operation Structure Chart of System Description of Algorithms Flowcharts
Troubleshooting	Explanation of How to Troubleshoot Product Chart of symptoms and possible causes Sample hardware and software test-data readings
References	Documents Related to the Product
Appendix	Specifications Schematic Diagram Component Layout Jumper and Switch Index Parts List Data on LSI Devices Program Listings

essary to install, operate, understand, and troubleshoot the product. This information comes from the laboratory notebook and related material. If the lab book was written with short summaries at the end of each major task, then they can be used in the technical manual with only a little revision.

When you write the technical manual, design it so that the reader can use it efficiently. Your purpose in writing the report is to provide the reader with enough information to set up, use, and fix your product. If the material is arranged logically and facts can be found easily, then your manual will be effective and helpful. Remember, your reader probably has no previous experience with your product, and those points that are obvious to you might be quite obscure to someone else.

In Table 4.2, the *introduction* is intended to clarify the rationale behind your product. It presents the nature of the problem addressed and how your product solves the problem. Any relevant background information and how your product relates to it should be discussed. Give a brief description of your product including its features and limitations. Your introduction should provide enough information so that a reader can determine if your product will solve his or her particular need.

The *installation* section should provide all the information needed to connect and test your product for proper operation. Sketches are useful to illustrate the equipment setup and to show the settings for any switches and jumpers. Consider the possibility of putting in a very brief "hook-it-up-quickly" section; this is for the reader who skips the instructions unless the product fails to work as expected.

A special section of the installation instructions should cover system checkout. Illustrate several different software and hardware configurations and how the equipment responds for each setup. If the installer has been having difficulty, an indication of correct performance will be quite valuable.

The *operation* section covers all the details of how to use your product. You might find a tutorial section and a reference section helpful in explaining the operation of your system. The tutorial section will aid the inexperienced user by illustrating some practical instructions on each of the product features; the reference section will help the knowledgeable user find needed information quickly. If you are doing a major product, the tutorial and reference sections may be easily separated from the technical manual and used alone by a nontechnical operator. Include a section on how to resolve operating difficulties. For example, if the user makes a single incorrect datum entry, explain how it can be corrected without reentering all the data.

The *circuit description* section explains all the technical aspects of your hardware. After an overview of the product, present a block diagram of the system and its division into various functional modules. Show each of the modules individually and explain how each operates. If appropriate, include timing diagrams and segments of the circuit diagram to help the reader understand the design. By explaining the hardware design in detail, the technical reader can repair the equipment himself rather than send it back to the manufacturer if service is required.

The *software description* presents the system programs and how they work together with the hardware. A program structure chart, or a flowchart equivalent, can help the reader understand system functions easily. Include a description of the algorithms and their flowcharts as necessary. By providing these details, the user can correct any software bugs and keep the system running. Depending on the reader and the nature of the system, you might include program listings in the appendix of your manual.

The *troubleshooting* portion of your technical manual is also intended to help the technical reader repair your product. The most helpful information you can give is a chart describing various symptoms of hardware and software malfunctions. For each of the symptoms, give a list of several possible causes and how to fix them. Keep it brief: a checklist at the workbench is far more valuable than page upon page of theory. Show several test setups, and give a number of typical voltage and oscilloscope patterns at critical points.

Much of the troubleshooting outline can come from your own experience in getting the prototype working properly. You probably already know which parts are critical and what the symptoms are if they fail. Additional information can be gathered for high-risk parts by replacing a good part with a defective one and noting the effect on the product. Open some critical circuits or cause some shorts in signal paths for additional troubleshooting data. If your system uses a microprocessor, include diagnostic programs to test various modules such as memory, I/O, and any external devices connected to the system.

The *references* section cites all the documents related to your product. The idea is for the reader to know what other material must be included with the design manual for a complete documentation package. You may also want to cite reference articles or tutorials for the reader to study.

TABLE 4.3 A SAMPLE OUTLINE OF A SOFTWARE DOCUMENTATION PACKAGE.

Problem Statement	Concise statement of the problem. Include a description of input variables required and outputs that will be provided by program.
Program Description	Overview: The approach to the problem's solution. Describe the strategy used to solve the problem. Include equations.
	Assumptions: What assumptions were made about the problem or the solution method?
	Variable List: Include list of names and descriptions of each of the input, process, and output variables.
	Data Structures: Sketch how the data is represented. This might be on several levels of abstraction: the problem level, the system level, and the machine level.
	Limitations: What parts of the problem are not programmed completely? How does the program handle undesired events such as out-of-range data or system faults? Areas that need more design in the future?
Program Design	Structure Chart: Describe the overall plan of how the program is constructed.
	Algorithms: Use pseudo-design language (PDL) to describe how the program works.
	Flow Chart: Illustrate portions of the program with a simple flow chart.
Program Listing	Provide a complete listing of the program. The program should be internally documented and start with a heading containing the name and function of the program, your name, and the date. Include comments to explain each major section of code. Tell how to compile and link all the program modules together.
Test Data and Results	Provide sample output that will illustrate correct operation of the program for normal and abnormal inputs. Include test data verifying the program at its design limits.

The *appendix* includes all the charts, tables, diagrams, programs, and background material related to your design. It is a collection of vital information required to describe and completely build the product.

When you assemble the technical manual, you can easily delete the software details and put them in a separate document. An outline of a typical software documentation package is shown in Table 4.3. For many products, it is convenient to give a brief explanation of the software in general and then refer the reader to a software manual. Because software tends to be updated with new revisions more frequently than hardware, a separate manual is easier to keep in order. This is especially true in a large project where different groups of designers are responsible for hardware and software.

4.3 DRAWING GUIDELINES

Drawing guidelines help you present your design clearly and consistently in a form other technical people will easily understand. As you saw in Table 4.2, a number of different drawings and tables are used to document a project. One of the most important drawings is the schematic diagram. The schematic diagram illustrates how your circuit is wired. It shows the logic gates, switches, connectors, memory, and all other devices that are connected in your design. Appendix A gives the guidelines to use when drawing the schematic.

Note that the schematic is oriented toward board-level rather than system-level documentation. For example, suppose you have a system with several printed circuit boards (PCBs) connected together. Each of the PCBs would be documented with schematics as in Appendix A. The system-level documentation would involve a different schematic showing the wiring that connects the PCBs to each other and to any external devices such as switches, lights, or power supplies. It would show the PCBs as functional blocks rather than display the details of each board.

The component-layout drawing is used with the schematic diagram to pinpoint where each part is located physically on the PCB. It should show the location of each IC and indicate all switch, jumper, and connector locations. If there are many switches and jumpers, a duplicate component-layout drawing with highlighted switch and jumper locations would be useful. The drawing should be drawn reasonably close to scale so that it appears the same as the board. Each part should be designated by its part number (U1, U2, etc.) and type (7400, 7404, etc.) for easy use at the workbench.

4.4 EXAMPLE TECHNICAL MANUAL

An example technical manual will be helpful in pulling together the ideas of this chapter. For example, consider the temperature monitor described in the Chapter 1 mini-proposal. How should this project be documented? One approach using Table 4.2 is shown in Appendix B. Because of the project's small size, not all the elements of Table 4.2 are present; however, you can install, operate, understand, and troubleshoot using the information given.

4.5 SUMMARY

All of your engineering work requires documentation of some kind to record your progress and to explain your final results. This is important because others need to understand and use your design or product.

Your primary documentation is the laboratory notebook. It contains everything related to the project: the initial ideas on a problem solution, the rough mini-proposal, the block diagrams, the hardware and software designs, and all your sketches and notes. Keeping the lab book well organized is important, and Table 4.1 outlines an approach for performing the lab work as a series of experiments. The lab notebook provides all the information you need for any reports on your project. It also contains the data to write a complete technical manual on the product.

A technical manual is one way of preparing a final report on a project. Its focus is on the user of the product and it illustrates setting up, operating, and troubleshooting the product. The technical manual can be organized in a variety of ways. Table 4.2 shows an outline of the major sections in one form of a technical manual. Several parts of the manual can easily be written as separate documents. For example, the operating instructions can be written in a small booklet kept with the product. Similarly, because of its many possible revisions, software documentation may be kept in its own separate binder.

All the schematics in the technical manual should be prepared according to drawing guidelines as shown in Appendix A. This is necessary because the design will be used by a number of technical people, and a standard format will not only help their understanding but also will reduce the chance of errors.

The example technical manual on the temperature monitor is intended to illustrate the documentation concepts in the chapter. Although abbreviated, this example is sufficient to help a user set up and use the equipment.

By systematically recording your daily work and documenting the results of your project, you will be able to explain your work clearly to others. In addition, you can review your progress and learn from your past mistakes. As you do the review, you may find a better direction for your work and perhaps even discover some new ideas for future product development.

FURTHER READING

BROWNING, CHRISTINE. *Guide to Effective Software Technical Writing*. Englewood Cliffs, NJ: Prentice-Hall, 1984.

FLETCHER, WILLIAM I. *An Engineering Approach to Digital Design*. Englewood Cliffs, NJ: Prentice-Hall, 1980. (TK 7868.D5F5)

SHERMAN, THEODORE A., and SIMON S. JOHNSON. *Modern Technical Writing*. 4th Edition. Englewood Cliffs, NJ: Prentice-Hall, 1983.

WEISS, EDMOND H. *The Writing System for Engineers and Scientists*. Englewood Cliffs, NJ: Prentice-Hall, 1982. (T11.W44)

FIVE

Clock Design Example

Reality

As you studied the first four chapters, your work was fairly structured and your designs were virtually guaranteed to perform properly. When you analyzed the customer needs and prepared a mini-proposal to outline your project plan, you were sure your plan would work. In the second and third chapters you used the proposal to complete a technical design on paper and to build a working prototype; you were sure this would work too. Why? Because you were both the customer and the designer: you could change the rules. If a feature became particularly difficult, you eliminated it from the requirements and made sure of a satisfactory design. Likewise, although your design had to adhere to certain technical standards (as in Chapter 3), you did not have to design to meet a formal industry or military standard. Consequently, your designs could be done easily and successfully.

The reality is that you must do your engineering in accordance with a number of technical standards. Which standards apply in any given situation depends on the customer and his or her requirements. For example, suppose your customer has a computer system that conforms with IEEE Std-696 (S-100 bus) and needs an added function in the system. Your design must either comply with the standard or run the risk of causing system malfunctions or complete system failure. Suppose your customer is the military and you would like to supply a circuit to be added to a radio transmitter. To have even a slight hope of proper operation, you would have to design your circuit to meet all of many applicable military standards and specifications. If your design does not meet the requirements in all points, then there could be system failures with drastic consequences.

Even though standards tend to be confining at times, they do ease the design burden considerably. When you design to meet a standard, many technical details are clearly outlined and do not require extra design time on your part. For example, in IEEE Std-696, a standard 100-conductor bus is defined with certain signals on particular lines in the bus. Thus, you do not need to design a connector and the location of certain signals. In addition to the physical configuration, a bus protocol is specified that helps you communicate with other devices in the system. Therefore, you do not need to spend design time figuring out bus protocol. The design effort, while complex in a sense, does become easier when you can use the standard.

The primary purpose of this chapter is to illustrate the complete design sequence from need identification through documentation. The secondary purpose is to introduce the IEEE Std-696 and how you use it when you design a product. To apply your knowledge of planning and implementation, you will design a clock that meets the IEEE Std-696. As you work through the design you will begin to understand engineering more fully and also learn how to design using an industry standard.

After you work through the clock design example, you should be able to develop the more-complex design required for a 68000 microcomputer system. Also, you can use your clock design as a full-scale realistic design guide to help you meet the bus standard. After you finish this chapter, you will be able to construct a prototype circuit board and document it fully. This chapter will help you integrate all you know so far and prepare you for more complex design projects.

5.1 NEED IDENTIFICATION

For our purposes, suppose you are a design engineer in a small design and manufacturing company and that you work directly with customers. A customer will normally meet with the sales and engineering managers first, and then you will be assigned to handle all the details of the project. When you first meet the customer, about all you can be sure of is that your company management thinks you can solve the customer's problem. Your role is to find out what the customer needs and then work out a solution to the problem. Although the customer will probably pay for a portion of your engineering time, assume your company expects to profit by building and selling several hundred units of the circuit you design.

On first meeting and talking with your customer, you find that he needs a real-time clock that will operate in an S-100 (IEEE Std-696) computer system running at the maximum system clock speed of 6 MHz. The clock should be able to keep time and date as well as provide a means of interrupting the system on either a regular basis or at a predetermined point in time. It should maintain an accuracy within ten seconds per month with or without system power. Except for initially setting the time and date, the operator should not have to interact with the clock in any way.

The clock should operate as a slave on the bus without affecting normal system operation; likewise, there should be no changes required in the system software. Data transfer between the clock and system should use I/O-mapped ports. The handshaking should use the S-100 RDY line, and because the processor operates at 6 MHz, some wait states may be used if necessary. Clock interrupts to the system should be switchable to use any of the S-100 vectored-interrupt lines.

Although the customer is planning to write his initial programs to poll the clock for the time, the clock should be able to use any of the vectored-interrupt lines to implement a future real-time multitasking executive. In the long term, this could involve multiprocessing with a new 6 MHz 68000 CPU board in the system. For the near term, however, the software will only set and display the time and service interrupts.

During these discussions with the customer, you discovered that he was really looking for two things: first, a clock in his system so that he could have a "time" function in his programs, and second, a future capability to use the clock interrupts in an advanced system being planned. If your clock design is a success, then there is a high probability that your company will have the contract to develop the advanced system as well.

5.2 PROJECT PLAN

At this point, you have a good idea of what your customer wants, and already you have a few ideas on how you can do the clock design. Before going out to the lab bench to build some models, plan your project! You have a job to do, so make a mini-proposal for your own use to keep you on target as you do the project. Put this mini-proposal directly into your lab notebook as you start work.

First, write your project definition as you understand it. Capture the essence of the big picture:

Project Definition

The goal of my project is to design, build, and test a prototype real-time clock board meeting the IEEE Std-696.

Next, write your project objectives. These are the specific measurable outcomes of what you intend to accomplish by the end of your project. Allowing at least a page or more in your lab book, you can write these objectives as your product specifications.

Project Objectives

Time: hour, minute, second
Calendar: day of week, date, month, year
Periodic interrupts: 100 ms and 1 s
Wake-up interrupts at preset times
Accuracy 0.01 %
Operating temperature 20 to 40°C
Power supply: +8 V (200 mA, rechargeable battery)
Meet IEEE Std-696
Use readily available parts, easy to build and test
No system hardware or software changes required

Once you have your project defined and have a set of objectives written in your lab book, figure out the strategy of how you can do the job. Avoid getting into the details of how to reach objectives, even if you think you know the answers. When you write the strategy, double space and allow yourself a full page in your lab book. Remember, this is a working document that you will want to refer to as you go along.

Strategy

The strategy of how to meet the clock objectives is to do a complete design on paper of the entire clock. This will be followed by building a prototype and testing it as each module is completed.

Finally, make your plan of action to get the project accomplished. Go through your strategy and find major items that need to be done and write yourself a ''do-it'' list. Refer to Chapter 1 and outline a step-by-step plan in your lab book along with your best estimate of how long each step will take. Sketch a tentative bar chart of the various tasks so you have some deadlines to meet.

Notice the sequence you followed when you started the clock project. After identifying the customer's needs, you wrote a project plan in your lab book that contained:

- Project Definition (the goal)
- Project Objectives (requirements and constraints)
- Strategy (how to reach objectives)
- Plan (action needed to implement strategy)

Now that you know where you should go, you can begin the implementation of the project. Remember that doing the planning as you start the project might seem like lost time, but in reality you are saving time by focusing your overall effort directly on the problem.

5.3 PROJECT IMPLEMENTATION

As you learned in Chapter 2, the project implementation involves two major steps: the technical design and the construction of a prototype model of the circuit. The technical design phase is when you do a complete paper design of the project; the construction phase is when you build a first unit and test it for proper operation.

5.3.1 Technical Design

When you begin the technical design, you should analyze the constraints and the specifications you intend to meet. Review the standards that you expect to use in your design and see that you understand what will be required; be sure to work from the standard itself and not someone else's summary of it. You want to synthesize a design concept that is a best balance between the required specifications and the various constraints.

First, analyze the product specifications in detail and categorize each requirement by function. Try to describe what your clock board is supposed to do; when you do this, avoid any statement about how to do it. You might use this list when you divide up the product by function:

1. What does the clock board do?
2. How well must it perform?

3. What system interactions are required?
4. What operator attention is needed?
5. What hardware interface is required?

After you complete answering these questions, probably you will have a functional specification similar to Figure 5.1. This specification describes the product as you see it at the moment. As you synthesize a solution to the design problem, you will add more technical details later that describe the actual circuit as it evolves.

Functional Requirement	Keep time: hour, minute, second Keep calendar: day, date, month, year Provide periodic interrupts to system Provide wake-up interrupts to system
Performance Requirement	Maintain 0.01% accuracy Operate between 20 and 40°C Use less than 200 mA at 8 V supply
System Interaction	Operate with 6 MHz system bus Assert interrupts when required Set interrupt modes
Operator Interaction	Set initial time and date Inquire for time and date Set interrupt modes
Hardware Interface	Compatible with IEEE Std-696 Operate as slave on the bus

Figure 5.1 The initial functional specification of the clock board. More details will be added as the design develops.

Next, sketch a system block diagram and then do a top-down design of the modules within the system. You might begin with a sketch of the system as shown in Figure 5.2(a). After seeing the board in its proper frame of reference, go into more detail as in Figure 5.2(b). Ask yourself what the board needs for inputs (power, someone to set the time at least once, etc.) and what it provides as outputs (time, interrupts, etc.). Note that these inputs and outputs match the functions you described in the functional specification in Figure 5.1.

At this point you can combine both sketches in Figure 5.2 to begin the block diagram of the clock board as shown in Figure 5.3. Now you are synthesizing a concept for the solution to the design problem. One possible concept is to use a CMOS LSI clock such as the National MM58167A real-time clock. The advantage of using CMOS is that you can operate the clock with a small battery over extended periods of time.

The block diagram for the system designed around this clock IC is shown in Figure 5.4. The design using this clock IC can be partitioned into several major modules that include the clock, address decoder, data-bus interface, interrupt switches, and power supply. Once divided into modules, the hardware can be easily designed in much the same way software is developed; that is, the hardware design can be done top-down, bottom-up, or most critical first.

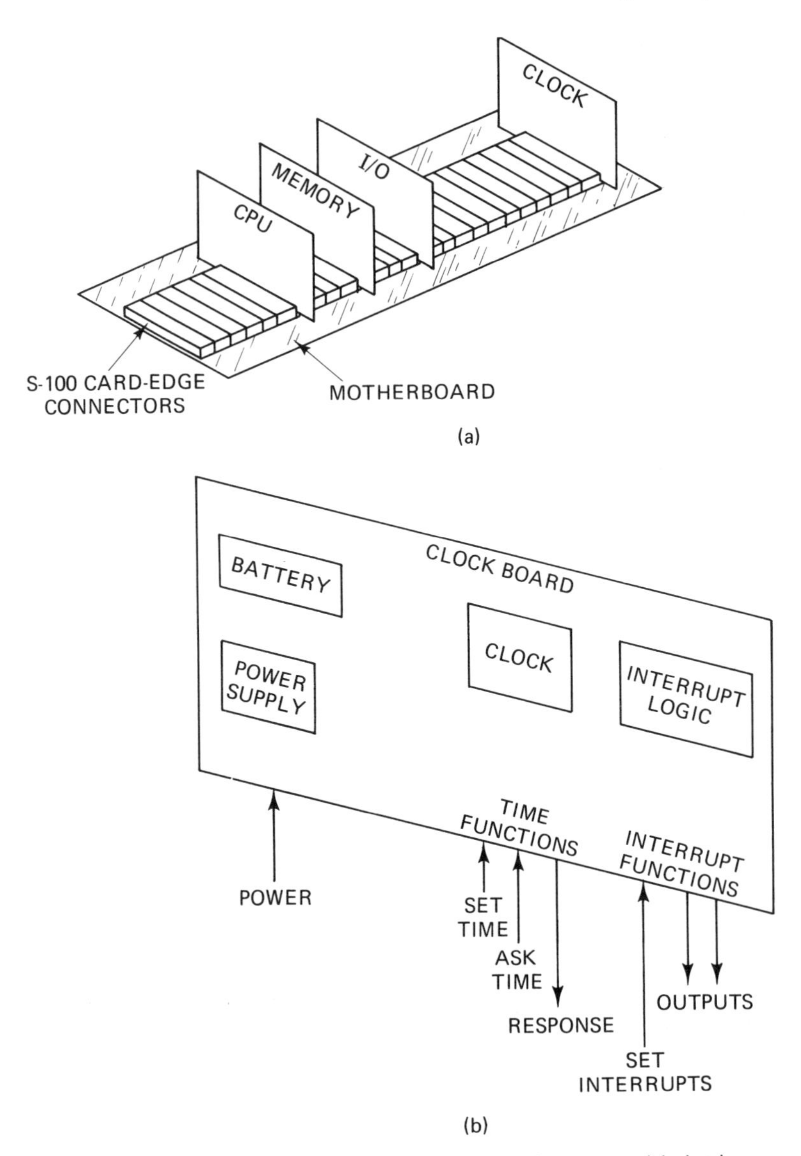

Figure 5.2 One way to begin the technical design is to draw some quick sketches. (a) Show the proposed clock board in the system. (b) Sketch the clock board showing its major inputs, outputs, and functions.

The appropriate design method in this case is to consider the most critical section first. Everything depends on the clock IC, and its requirements should be considered before all else. The block diagram of the MM58167A, Figure 5.5, shows that the IC uses 5 address lines (32 different addresses) and a chip select, has a single bidirectional data bus and separate read/write controls, two interrupt outputs, and a power-down control. Each

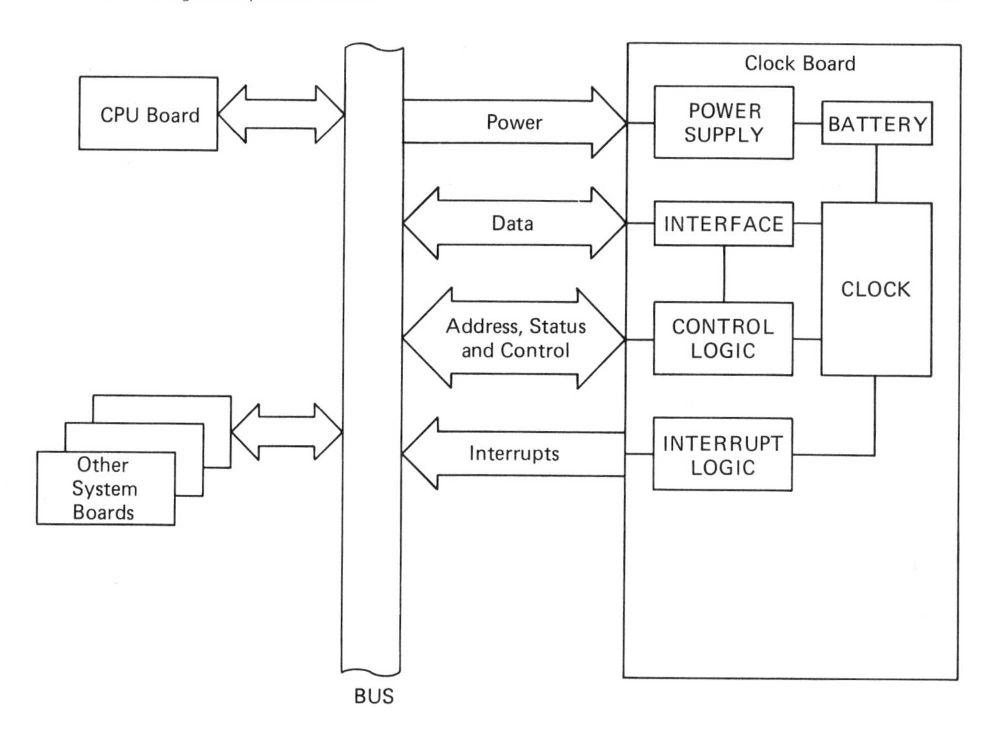

Figure 5.3 Block diagram of the system plus some rough detail on the clock board design.

of these requirements can be designed individually at this point and result in the implementation shown in the circuit schematic in Figure 5.6.

When you use I/O mapping (as opposed to memory mapping) for the clock data transfer, you have a maximum of 256 ports. Only the lower eight bits of the address bus need decoding, and five of these are internally decoded by the clock IC. For maximum flexibility, you can use DIP switches to select the three most significant bits of the I/O address. The output of the decoder is used as a chip-select for the clock and is used by the read/write qualifier circuit.

The S-100 data bus in-and-out lines are buffered with a pair of 74LS244s connected directly to the clock IC. The buffer for data-in (DI from the processor's viewpoint) is strobed using the I/O read qualifiers pDBIN, sINP, and the chip-select CS. The data-out buffer is strobed using pWR*, sOUT, and the chip-select CS. During a typical read bus cycle, for example, the proper address is placed on the address bus, sINP asserted, and then pDBIN asserted to strobe the DI buffer and the clock RD* input. In the typical write bus cycle, the address is put on the bus, sOUT asserted, and then pWR* asserted to strobe the DO buffer and the clock WR* input.

Timing. A basic read or write bus cycle has three bus states (BS1, BS2, and BS3), each of which takes 167 ns in a 6 MHz system; consequently, a bus read or write

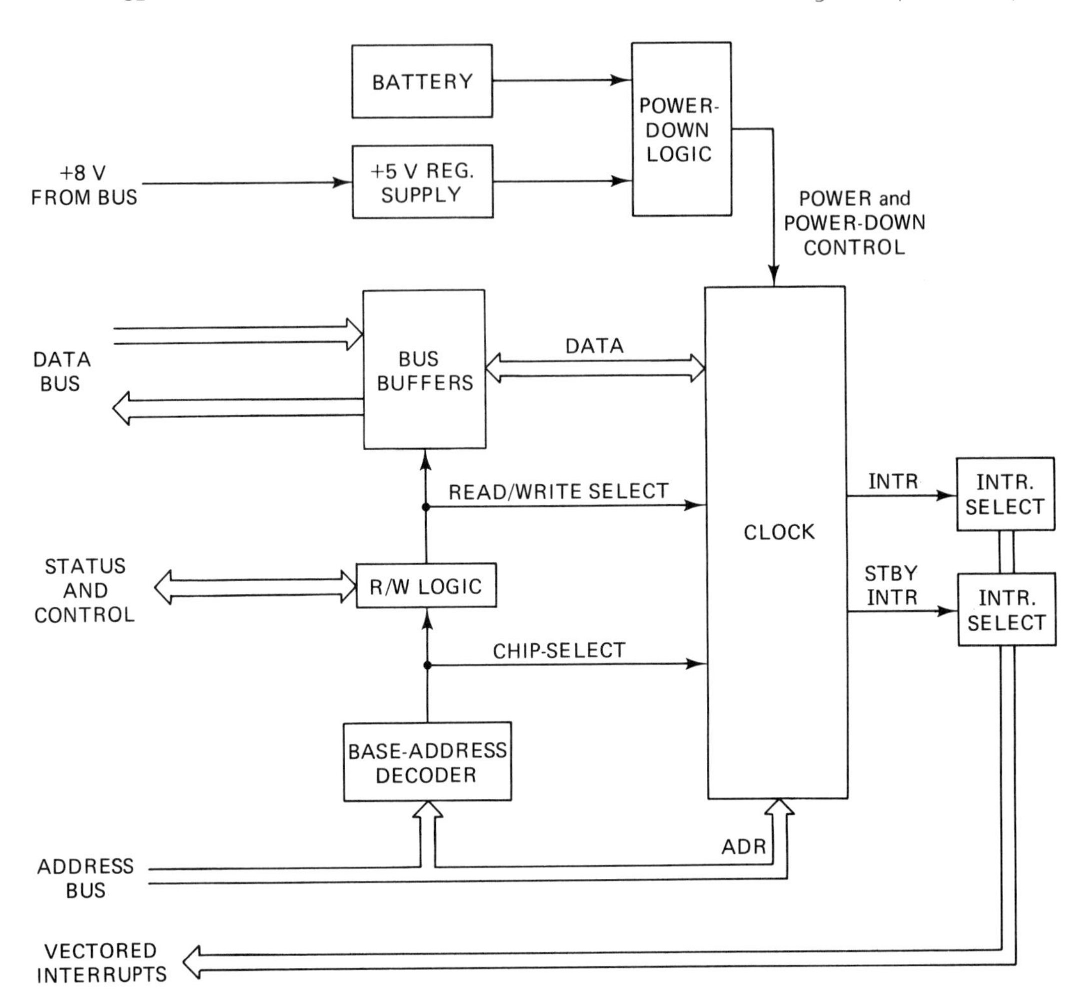

Figure 5.4 Block diagram of real-time clock board.

normally completes within 500 ns. However, for I/O devices that require substantial access times, the bus cycle can be extended by adding a number of wait states, as in Figure 5.7. To find out if wait states are required during any read or write bus cycle, the bus master (the CPU) samples the RDY line at the rising edge of the system clock in Bus State 2 (BS2). If the RDY line is found low, then a wait state (BSw) is inserted immediately after BS2; if the RDY line is still low a bus state later, then yet another wait state is inserted. Wait states continue to be added until RDY finally goes high; at that point, the bus cycle concludes with BS3.

The MM58167A clock requires wait states in the bus cycle because of its slow access time. For a clock-read operation, the time required from a valid address until the output data is valid might range from 500 ns to the specified maximum of 1050 ns, far

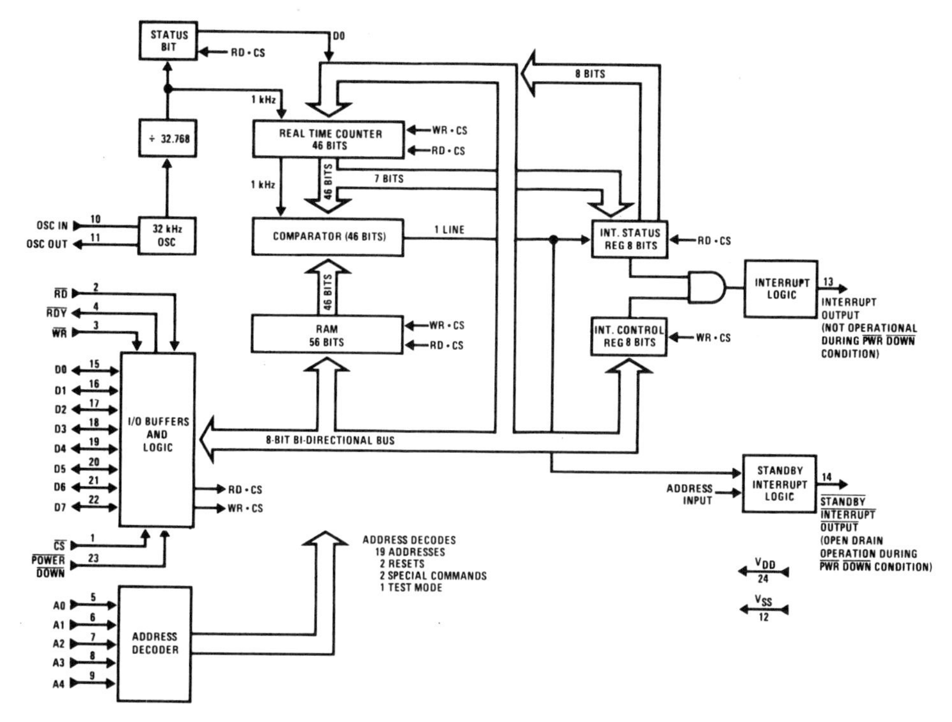

Figure 5.5 Block diagram of the MM58167A real-time clock. (Courtesy National Semiconductor Corporation)

longer than the time required for a normal bus cycle (i.e., 167 ns $\times$ 3 = 500 ns). The read cycle of a typical 6 MHz IEEE-696 system is shown in Figure 5.7, with approximate times to scale. Notice that the example timing diagram includes a total of three wait states to make up for the slow access time.

The normal S-100 requirement is that the RDY line be low at the end of BS1 if waits are required in a bus cycle. In the clock case, however, this is impossible: the clock is not even being read until BS2, so it cannot respond until about a cycle later. In a case like this, the CPU board itself has to add in a wait state automatically if an I/O access is being made. Doing so allows the clock to have until the end of BS2 to get RDY pulled low. As shown in Figure 5.7, the clock holds RDY low to get two more wait states in the bus cycle. When the clock data is finally valid, the RDY line is allowed to go high, and the data is read by the processor in BS3. To finish BS3, pDBIN is negated, which raises the clock RD* input, and the address deselects the clock.

Figure 5.8 shows actual bus timing data using a Z-80 CPU board as bus master. The CPU has been set to add in a single wait state. RDY is pulled low by the clock about 50 ns after the start of BS2; this is 167 minus 50 or about 120 ns before it *must* be low. The IEEE Std-696 calls for a 70 ns setup time before the rising edge of the next system clock, so you can see there is about 120 minus 70 or 50 ns margin here. That is, if RDY were later by more than 50 ns, then RDY would not meet the standard's setup-time

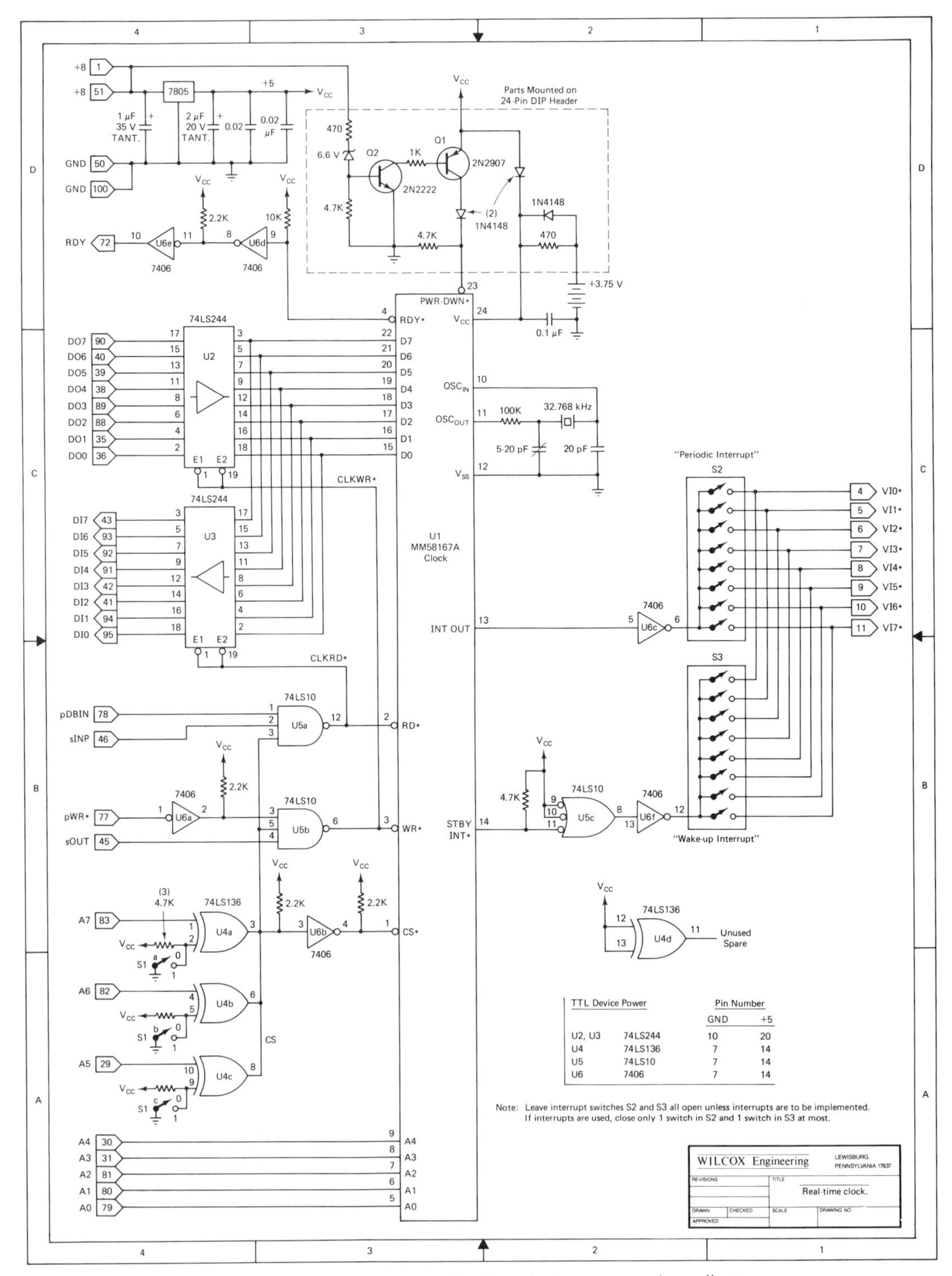

TTL Device Power		Pin Number GND	+5
U2, U3	74LS244	10	20
U4	74LS136	7	14
U5	74LS10	7	14
U6	7406	7	14

Figure 5.6 Real-time clock schematic drawn on zonal-coordinate paper.

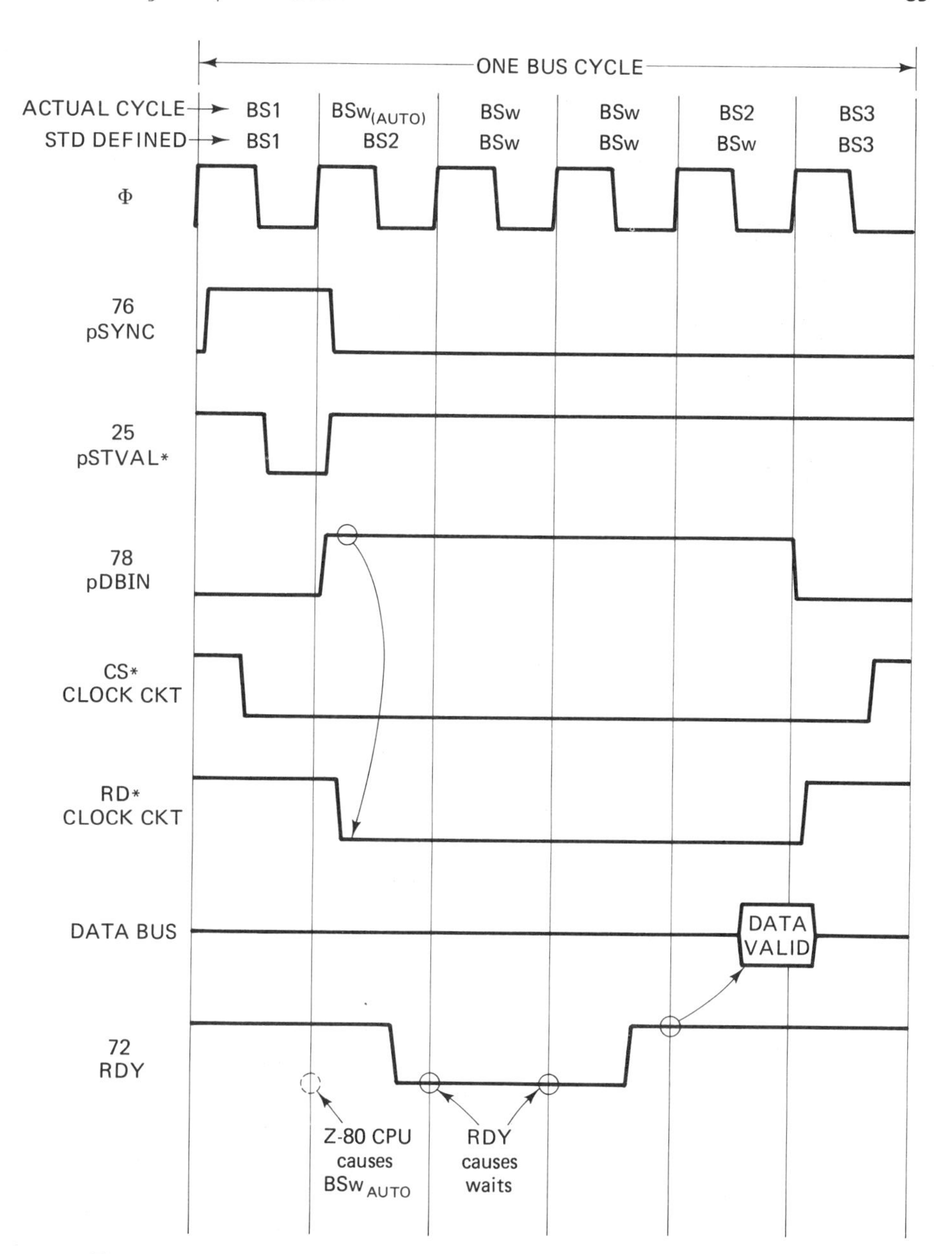

Figure 5.7 Read bus cycle for the clock. As drawn, one wait is caused by the Z-80 CPU board; the others by the clock RDY line.

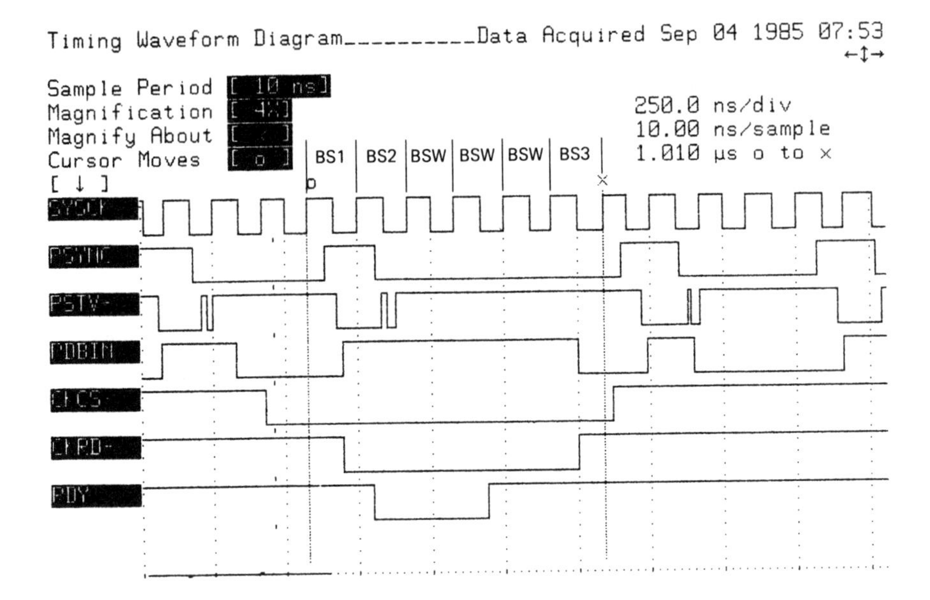

Figure 5.8 Actual bus data using a commercial Z-80 CPU board.

specification. Your system would probably work even if the setup time were reduced to, say, 20 ns, but you would not be able to guarantee a general-case performance.

The write bus cycle timing requirements for the clock are quite similar to the read later by more than 50 ns, then RDY would not meet the standard's setup-time timing described above. As in the read, a total of two wait states must be included to allow sufficient time for RDY to be pulled low during the write operation.

Power-down design. The power-down design is quite critical in the actual operation of the clock. If PWR-DWN* is not asserted to a low-voltage level at least a microsecond before the system power is removed, then the contents of the clock memory might be anything (or nothing) when normal system power is restored. Further, when the power does come back on, PWR-DWN* must be held LOW until all bus signals are valid.

A common way of accomplishing this is to use a zener diode to set a reference threshold voltage to control a transistor switch. Consider the circuit, Figure 5.6, when the system is off: Q1 and Q2 are both off. When the system power comes on, nothing happens until the system bus voltage reaches about 6.6 V. Then the zener conducts, turns on Q2, which then turns on Q1. When Q1 goes on, voltage is applied to PWR-DWN*, which activates the clock chip for normal operation. A system bus voltage greater than 6.6 V is enough to allow near-normal output from all 5 V regulators, so all bus signals should be valid by this time. When the system is turned off, the system voltage drops down gradually enough so that when it passes below 6.6 V, the zener and Q1/Q2 can drop the PWR-DWN* to a low voltage before the bus signals b ecome invalid.

A small 3.75 V, 20 mAH NiCad rechargeable battery is used to maintain clock operation when system power is off. Specified current consumption of the clock is on the order of 10 to 20 μA using the battery with series-blocking diode. The 470 Ω resistor provides a trickle charge of 0.5 mA to about 2 mA depending on the state of the battery and component tolerances. For an average of several hours a day use, this resistor keeps the battery charged between 3.8 V and 4.15 V with no difficulties.

Logic circuits. IEEE Std-696 (Sec. 3.7) states that a card may not source more than 0.5 mA at 0.5 V nor sink more than 80 μA at 2.4 V on most system signal lines. The effect of this is that only one LS-TTL gate per card can be connected to each line of the address bus. The address decoder uses a 74LS136 exclusive-or, which sources a worst-case maximum of 0.8 mA at 0.4 V. This should cause no operational problems in any practical sense, and the appropriate design tradeoff is to use the 74LS136 rather than add extra hardware to buffer the address bus to be strictly within the specification. You might want to check with your customer to verify that this tradeoff will not cause a problem later in the system.

The pull-up resistors for the open-collector 7406 are selected to maintain proper logic levels and currents. The design tradeoff is to let the various resistors be high for lower current consumption, or to let them be low for more speed at the expense of the current. As drawn in Figure 5.6, the pull-up resistors are designed for as much speed as possible.

The clock standby-interrupt output can be used to implement the "wake-up" interrupt specification; to do this, an extra noninverting gate is needed. The spare 74LS10 and the last spare 7406 gate can be used to get this output from the clock board without adding another package. This particular output does not have the flexibility of the programmable output, but is useful to indicate a match between real time and a preset time for a wake-up alarm.

5.3.2 Prototype Construction

The second half of the project implementation is the construction of the prototype model of the circuit. Although the paper design is done and appears correct, you still must build the prototype and demonstrate its proper operation. In the technical design, it is easy to overlook some specifications, and your prototype will help you find and correct these oversights.

First do a rough sketch of where the major functions will be physically located on the prototype circuit board. Then build the prototype module by module, much as you did the technical design by modules or functions. In the design phase, the most critical section was the clock, and you did its circuit first. After the clock, you did the address decoding, bus control, interrupts, and power-down. When you build, however, construct from the bottom-up so you can debug as you go along.

What you would like to have as you build the prototype is a working, error-free circuit that can help you catch flaws in your paper design. Start with a simple but necessary function on the clock board: the power supply. All the parts on the board need power, so build that module first and check it out. Once you have power available, then you can do the address decoder and check it out too. Next, do the bus buffers and continue with the rest until you have a complete circuit.

How do you do the testing and then the debugging if there is a problem? Test the first module, the power supply, with a multimeter for shorts when you finish its wiring. Then plug the board into an S-100 system and turn on the system power. Measure the +5 V from the regulator and see that it appears at the proper places on your prototype board. If you find a problem with your supply, trace the circuit back to see that the system +8 V is coming into your circuit. If it is, with only the 7805 regulator in the supply, then either the 7805 is defective or its output is shorted. You should have found the short earlier with the multimeter, so fix the regulator. When all is satisfactory, turn off the system, remove the board, and start wiring the next section.

Build and test the address decoder next. Wire up the four logic gates that produce the clock chip-select CS*. Set the three switches to a clear I/O port address (A0 hex, for example) and plug the board back into the system. Apply power and boot up your system as you do normally. If your system cannot boot, the problem must be related to the A7 through A0 address inputs you connected on the clock board. Use your logic probe and check to see if one or more of the address lines is being held high or low. If the system booted normally, use the logic probe and check for CS*. You should find activity here because the address bus hits the address A0 (and all the combinations of 101x xxxx) many times each second.

Following the same approach, build and test the read-and-write strobe qualifier circuits RD* and WR*. Boot your operating system and use the logic probe to make sure CS* still has activity. Check RD* and WR*: they should be negated (i.e., HIGH). Using your operating system debugging program (for example, CP/M's DDT), execute a small one-line program that will assert RD*. If your system is running with a Z-80, you might execute an instruction

```
IN A, (A0)
```

This will cause the Z-80 to place A0 on the address bus and then read the data at that I/O port into the accumulator. Relate your circuit to the read bus cycle in Figure 5.7: the combination of pDBIN and sINP and the correct address will cause RD* to go LOW just once. You should see this pulse on your logic probe as you execute the instruction. You can do a similar test for WR* by executing the instruction

```
OUT (A0),A
```

This will cause the Z-80 to send the accumulator data out to the I/O port A0. Each time you execute this instruction you should see the pulse on your logic probe.

If you already have an IEEE Std-696 68000 CPU board in operation, you know that it does memory-mapped I/O. That is, rather than IN or OUT, the 68000 normally exchanges data with other system devices by reading or writing to memory addresses. A certain block of the 68000's 16 Mb address space should be decoded on your 68000 board for I/O purposes. For example, suppose the address range FF0000 to FFFFFF hex is designated for I/O devices. You could expect that the S-100 sINP and sOUT would respond properly and you could write your test code just as you would for memory accesses. For example, you might use

```
MOVE.B $FF00A0,D0
```

to do a read from port A0. The data would be stored in D0. Then use

```
MOVE.B D0,$FF00A0
```

to write D0 data out to port A0.

After you have RD* and WR* working properly, you can build and test the data bus buffers. With nothing in the clock socket, you will not be able to read or write actual data. You will, however, be able to boot your system and see that the buffers are being properly strobed when you execute the same programs you used to test RD* and WR*. If you want, you can test for proper reading by temporarily connecting arbitrary logic HIGH and LOW values to the data lines at the empty clock socket: execute an IN A,(A0) and see if you read the correct data into the Z-80 A register. Remove the temporary wires and execute an OUT (A0),A instruction to check if data is getting to the clock socket. When you do this instruction, you should be able to use your logic probe to see a pulse at each data pin on the socket.

Finish the rest of the clock board following the same approach: build a module, test it, build another module, and test it too. When you have the board finished, you will fully understand how it works and will be confident of how well it meets the specifications. Notice that the only test equipment you needed to get the clock board in operation was your multimeter and logic probe. You might have used an oscilloscope and executed scope-loop programs to check the various timing waveforms and be sure they were within specifications.

For illustration in this chapter, the entire real-time clock was prototyped on a Vector 8800V wire-wrap board as shown in Figure 5.9. The interrupt switches (S3) for the wake-up interrupt are located beside the main interrupt switches (S2). Less than a third of a standard S-100 board is required for all the circuits. The transistor power-down circuit is wired on the 24-pin component-carrier at the lower-left edge of the board by the battery. Plastic ''wrap-ID'' panels were used on the back of the board for each IC socket to make component and pin identification easier.

5.4 DOCUMENTATION

When you started the clock project, you began by writing your project definition, objectives, and strategy in your lab notebook. After writing out a step-by-step plan of action, you did all your circuit design in the notebook. Then, as you started building and testing the prototype, you put more information in your lab book. By the time you had a working prototype meeting specifications, your lab book probably became quite full with everything in it. Of all the information, the most important is your test data and your notes on how you corrected any problems.

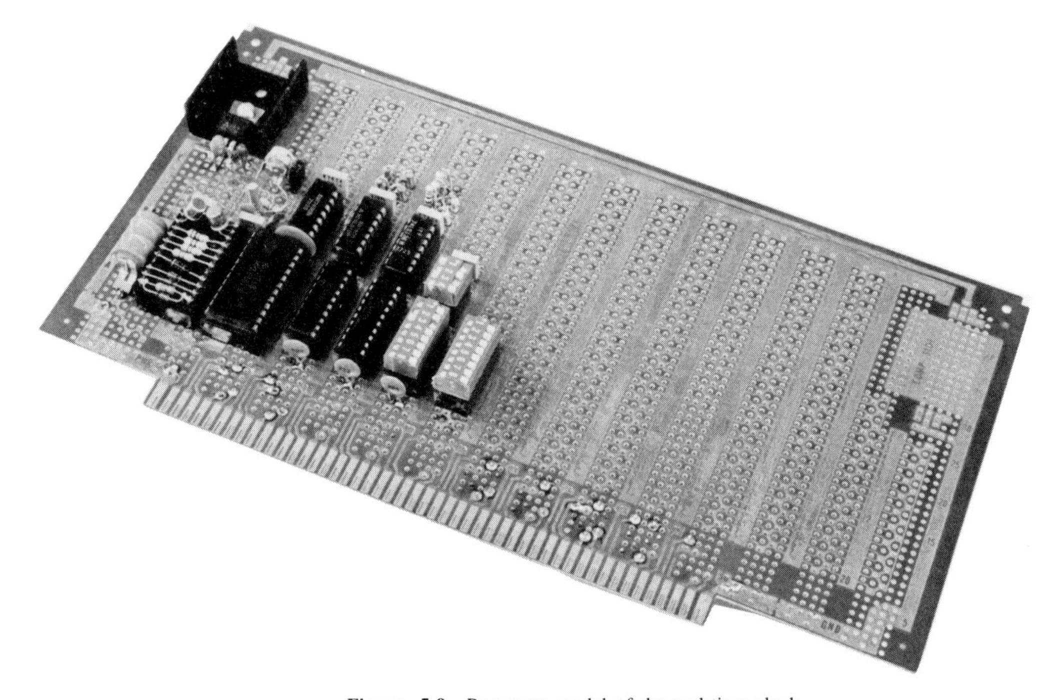

Figure 5.9 Prototype model of the real-time clock.

What you do next really depends on your actual situation. If your customer (hypothetically described earlier in Section 5.1) wants to try the prototype in his system so he can write some software for it, then all you may need to give him is a single sheet showing the address and interrupt switch settings. More likely, however, you will need to provide more information than that.

Suppose you learn that the design is entirely satisfactory and that your company will probably sell several hundred of the clock boards. The information in your lab book will be necessary so you can provide sketches for your drafting personnel and a parts list for purchasing. As your board finally moves into manufacturing, you will need to verify production test data with your original design. If you were diligent in keeping your lab book up to date, then you should have no problems.

A manual of some type will be shipped with the clock boards you sell. For the moment, assume that a full technical manual is required. Refer to Table 4.2 and start assembling your information. All the data you need should be in your lab book and only needs to be organized and explained.

5.5 SUMMARY

Your work during the first four chapters was structured enough so that your designs would perform properly. Other than making sure that your circuits were technically correct, you did not need to follow any particular industry or military standards. In reality, though, you know that you must not only design a technically correct product that satisfies the customer's needs, but you must also design to standards.

Standards can substantially ease your design burden. The idea behind the standards is to have uniformity and interchangeability of system subassemblies, and this means that many technical details are already set. You do not need to spend time doing design that the standard has already covered. You do, however, have the obligation to study and understand the standard so that you can use it properly.

An ideal way to learn about a standard is to do a design with it. The purpose in doing the clock board project is to see how to do a complete design using the IEEE Std-696. This particular standard has been successfully implemented on a number of machines, and you can test out your circuit easily using one of them as a test bed.

Just as you would begin any design project, when you design with a standard you start with the need identification and project plan. Write down the project definition, objectives, and strategy in your lab notebook. Then move on to a complete paper design and the construction of the prototype.

Most of your work with the standard is during the technical design phase when you examine how your circuit should interface with the system. If you begin with a top-down approach of defining various functions that your board needs to perform, then you can quickly sketch a useful block diagram of the board. Deal with the most critical section as soon as you have a block diagram: examine the clock itself and what support circuits it needs. Then you can design the necessary interface modules to meet the standard.

The most critical issue when you design with the IEEE Std-696 is the timing of signals. Depending on the type of input or output instruction the CPU performs, the bus signals behave in a unique way to complete the I/O required. You must sketch your timing diagrams and examine them closely to be sure that your circuit will be able to respond properly. In the clock design, for example, you found that the clock board could not respond quickly and that wait states were necessary for proper operations. Without a timing diagram, the need for wait states would not have been clear at all.

When you construct your prototype clock, do it bottom-up so you can test and debug along the way. Build the power supply section first because all the modules depend on it for their power. After you finish the supply, build the address decoder and test it for proper operation. You can build and test the whole clock board this way so that when you test the last module, you are confident you have a satisfactory product.

Your lab notebook is the central focus of your project documentation. All your project plan notes and technical design work is recorded here. Perhaps most importantly, the data you took when you tested each prototype module is there along with notes on how you corrected problems you encountered during testing.

After the prototype is built and tested, you need to document your work for others involved in the project. Perhaps you need only make a brief summary of the design for use in your own department. Maybe you need to extract more information from your lab book and put together a full technical manual, as you learned in Chapter 4.

By now, you should be fully capable of developing the engineering design plan for the 68000 microcomputer system in the next chapter. Naturally, the 68000 project will be more complex, but your approach will make the difference. You know that you can plan, design, build, and test in small steps that will finally get the job done successfully.

EXERCISES

1. One of your first engineering assignments with your company is designing a clock to run with a Z-80. The customer would like you to design a small Z-80 system with a clock connected directly to it; no bus interface is required. Write a project plan. In your plan, include the project definition, objectives, strategy, and the plan for implementation. Make assumptions as needed.
2. Sketch a block diagram of the Z-80 and clock system. Hint: see the MM58167A data sheet for application notes.
3. Discuss the merits of I/O mapping versus memory-mapped addressing of the clock board. Make an engineering decision and explain your rationale.
4. Sketch the Z-80 read cycle timing diagram (memory or I/O depending on your choice in Problem 3). Sketch the MM58167 read cycle timing diagram aligned under it on the page.
5. Assume that the Z-80 runs at 2 MHz. Will it read the data correctly from the clock IC? If the Z-80 runs at 6 MHz instead of 2 MHz, will the data be correct? What must be done?
6. Do Problem 4 for the write cycle timing.
7. Do Problem 5 for the write cycle case.
8. Explain the steps you would take in constructing a prototype of the Z-80 with MM58167 clock.
9. How would you test the prototype?
10. Write a two-page technical manual describing your computer board with the clock connected to it.

FURTHER READING

CALAWAY, J.L., and B. HILL. "CP/M, Your Time Has Come." *BYTE* (May 1982): 479–93.

CIARCIA, STEVE. "Everyone Can Know the Real Time." *BYTE* (May 1982): 34–58.

HASSEBROCK, GLEN E., JR. "The Hayes Chronograph, an H-89, and CP/M." *Remark* 35 (Dec. 1982): 9–14.

IEEE Standard 696 Interface Devices. New York: IEEE, 1983.

LIBES, SOL, and MARK GARETZ. *Interfacing to S-100/IEEE-696 Microcomputers*. Berkeley, CA: Osborne/McGraw-Hill, 1981.

MM58167A Microprocessor Real Time Clock. National Semiconductor Data Sheet, July 1984.

POE, ELMER C., and JAMES C. GOODWIN. *The S-100 and Other Micro Buses*. Indianapolis, IN: Howard W. Sams & Co., 1981.

The TTL Data Book. Vol. 2. Dallas, TX: Texas Instruments, Inc., 1985.

WILCOX, ALAN D. "Designing a Real-Time Clock for the S-100 Bus." *Dr. Dobb's Journal*, 10 (7): July 1985, 56–90.

SIX

System Planning and Design

The Big Project

As you begin a big project, you can almost see your circuit board all wired and operating in a complete system. Then the doubts begin as you wonder if you know enough to do the design or even enough to build and test it. This is how it is with a challenging project: you are never quite sure of success until the design is finished and you see it actually working. The 68000 project is a big challenge and cannot be done easily, but it *can* be done.

This chapter shows how to plan a complete 68000 project so that it will be a success. The promise of Chapter 5 was that, after working through the clock design example, you would be able to develop the engineering design for a 68000. The key, of course, is the plan. You can use the clock design as an example while you make the 68000 plan. Just as you did before, start by identifying needs and then defining the project and its objectives. After you decide on a strategy, make your plan of action.

When you did the clock board, writing the project definition objectives and strategy probably seemed like simple busywork. Why bother writing it all down when you could keep it in your head? The plan of action with a task list and a schedule probably seemed the same. Why bother doing all that for such an easy plan? The answer is so that you would have some practice making a project plan and so that you would have an example to follow. You cannot finish the 68000 project without careful planning and without watching your schedule.

6.1 NEED IDENTIFICATION

Assume the scenario as in Chapter 5 and suppose that you are a design engineer in a small company. Your clock design was a success, and now your company has a contract to develop a 68000 computer system. Your role is to identify what the customer needs and then go ahead with the design and construction of the prototype computer board.

On meeting with your customer, you find that he wants a 68000 computer board that will operate on an S-100 bus (IEEE Std-696) at 6 MHz. At the present time, the customer has a 6 MHz Z-80 CPU board running under CP/M 2.2 and wants to upgrade the system. As you understand him, the purpose in the upgrade is so his simulation programs and other applications programs can run faster. If the new board has a substantial speed improvement over the old model, he believes he will be able to sell a number of them to his customers who are already using his Z-80 CPU boards.

An important factor in marketing the new boards is the issue of software compatibility. How can he convince his customers to buy a new 68000 board when they cannot use their existing software? Is there some way to run their existing Z-80 applications programs by using a slave Z-80 running as a task under the control of the new 68000 board? Then, over a period of months, the most critical programs could be converted to 68000 code for faster execution. This way he could help his customers phase in the 68000 board while at the same time avoid making their Z-80 programs obsolete.

To help you develop the new 68000 CPU board, your customer is willing to lend you one of his Z-80 systems. You will have, in addition to your normal lab equipment, an IEEE Std-696 system that has the S-100 motherboard and power supply, a disk-controller card, an I/O card, and a 64K static memory board. There are many empty slots in the cabinet for you to insert your new 68000 board when you have it ready to try out. If you need more memory, your customer has several more 64K static memory boards, all with extended-addressing capability.

6.2 PROJECT PLAN FOR THE 68000 CPU

Based on all that you know about your customer and his needs, you can now put together the project plan for the 68000. This plan takes the form of the mini-proposal that you learned about in Chapter 1 and used again in Chapter 5 with the clock project. The 68000 project plan you do in this chapter is just the plan itself; the implementation of the plan is described in the Appendix.

I. Project Definition.

The goal of this project is to design, build, and test a prototype 68000 CPU board that meets IEEE Std-696.

II. Project Objectives.

CPU software requirements:

Monitor in EPROM that will control the 68000. It should be able to read and change memory and registers as well as trace and execute programs. It should provide a means of downloading programs from a host system.

BASIC programming language in EPROM.

Disk operating system that will boot from a disk and provide normal system utilities and languages.

CPU hardware requirements:

- Use MC68000 microprocessor
- Be able to run using MC68010 microprocessor
- Conform to IEEE Std-696 (S-100 bus)
- On-board memory
  - Two 28-pin sockets for static RAM (6116 or 6264)
  - Two 28-pin sockets for EPROM (2716, 2732, 2764, or 27128)
- I/O data transfer by memory-mapping; I/O-mapping above FF0000 hex
- Clock speed 6 MHz
- Wait generator to select 1 to 8 wait states
- Reset circuit to provide stack pointer and program counter for 68000
- Bus-error watchdog timer circuit
- Build with easy-to-find parts

Testing and maintenance requirements:

- Single-step circuit with control switches on CPU board
- LED to indicate 68000-halted condition
- Switch to set on-board memory anywhere in 16 Mb range of 68000
- Freerun capability
- Self-testing capability

III. Strategy to Achieve Objectives.

The overall strategy will center on a sequence of design, build, and test; this sequence will be used to develop each module of the CPU until the prototype board is complete. To support this strategy, the 68000 will be configured in a freerunning mode until the hardware is ready for the monitor EPROM. The routines in the monitor EPROM will be used to aid the testing as more hardware modules are added to the board. Finally, the monitor EPROM will control the 68000 to load a disk operating system (DOS).

IV. Plan of Action.

Review background information:

- Review Motorola application notes and technical manuals
- Review trade and technical articles on 68000 design
- Study designs of existing 68000 boards
- Study designs of S-100 bus boards

Order prototyping parts:

- Prototype board or wire-wrap board

- Clock oscillator
- ICs not on hand already

Study IEEE Std-696 in detail:

- What signals are required of a bus master?
- How to derive the required signals?
- Synchronization of bus states with 68000 states

System design:

- Draw block diagram of system
- Consider hardware/software tradeoffs
- How does the bus interface standard affect this?

Hardware design:

- Identify critical modules
- Draw block diagrams of each module
- Do logic diagrams, timing diagrams
- Build and test each module

Software design:

- Top-level system design
- Memory map and how modules all fit together

Document the system (from lab book details)

Test the completed board in the system

Evaluate how well the board meets system specifications

6.3 GENERAL STRATEGY

If you have been following the approach described in the Chapter 5 clock design, you know that implementation of the plan is next. When you did the clock, you did a complete paper design before you started building and testing the prototype. For a project the size of the 68000, however, there are just too many unknowns that prevent doing a complete paper design before construction.

A reasonable solution to this problem is to design the system by modules and do the construction and testing of each module immediately. After you design a module, build it on a prototype board, test it as completely as possible, and move on to design another module.

This sounds easy enough, but there is a slight difficulty: how do you know which module comes next? In Chapter 2, the technique was to partition the system into modules by drawing a system block diagram; then, using a top-down approach, the system diagram could be divided into manageable modules. At this point in the 68000 project, you have information on the system and understand some of the requirements the CPU board has to meet. You can draw the system diagram along the lines of Figure 6.1. After you have the

Project Schedule

	Jan				Feb				Mar				Apr			
Week Beginning	6	13	20	27	3	10	17	24	3	10	17	24	7	14	21	28
Tasks to Do																
Review background info	* * * *	* * * *	* *													
Order prototype parts	* *	* *														
Study IEEE Std-696		* * * *	* * *	* * * *												
System design		* *	* * *	*												
Design, build, test			* *	* * * *	* * * *	* * * *	* * *	* * * *	* * * *	* * * *	* * *	* * * *	* *			
Software design					* * *	* * * *	* * *	* * * *	* * * *	* * * *	* * *	* * * *	* * * *	* * * *	*	
Document system											* *	* * * *	* * * *	* * * *	* * * *	* *
Test completed board													* *	* * * *		
System evaluation														* *	* * * *	* *

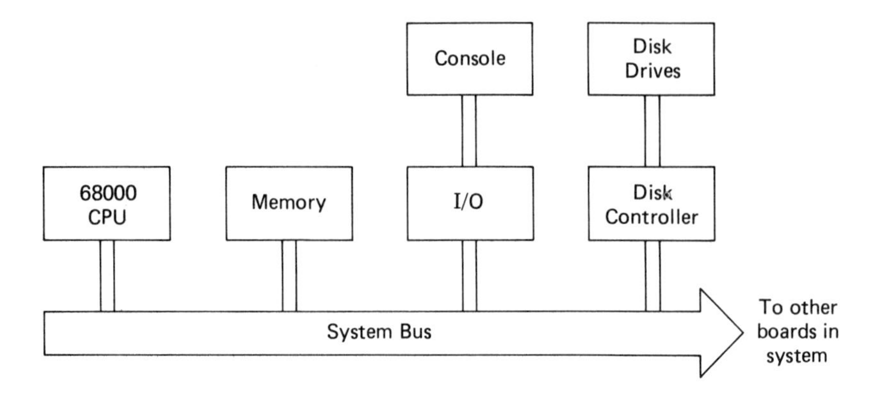

Figure 6.1 Block diagram of the computer system.

CPU board in perspective, review the specifications and go into more detail. A rough sketch of the board modules might look like Figure 6.2; this is not a complete block diagram yet, however.

If you use Figure 6.2, can you guess which module is first, second, third? Suppose you do the power first—an essential circuit and easy to design, build, and test. How can you design the power supply without knowing how heavily it will be loaded? Just make a reasonable guess and hope you were close enough? Probably this is the best choice for the prototype, because you do not know exact loading until the design is completed.

Consider designing and building the clock and the reset circuitry next. The 68000 has to have both of them, and they can be done quite simply. Do you see the essential strategy here? Design, build, and test—gradually building a foundation for more complicated modules that come later. Constantly check for proper operation of each module as you add it in; check the previous modules to be certain they still work.

Designing for testability means that you engineer a product so that it can be easily tested. If you were to do your design from start to finish on paper and build it all at once, you would probably have problems testing for proper operation. This is the situation a service technician might be in when troubleshooting a customer's board: how can it be tested easily? In contrast, if you do your design so that it can be built and tested by modules, chances are good that the final product will be easy to test and service. Why is that? Each of your modules was tested, and you provided for it in the initial design.

As an example of designing for testability, suppose you prepare a circuit to single-step the 68000. Normal system operation is for the 68000 to keep running; being stopped indicates some error, and the watchdog timer alerts the system. To avoid having the timer cause an alarm, you design your circuit so that the timer is disabled if you are in a single-step mode. If you were not building and testing during your design, you could overlook this simple timer-defeat logic.

In fact, your entire design is based on testability. The 68000 board can be built and tested with common lab equipment: multimeter, logic probe, and oscilloscope. If you did not build and test your modules as you went along, you would need more powerful tools such as a logic analyzer or an in-circuit emulator for success. You can see the parallel here

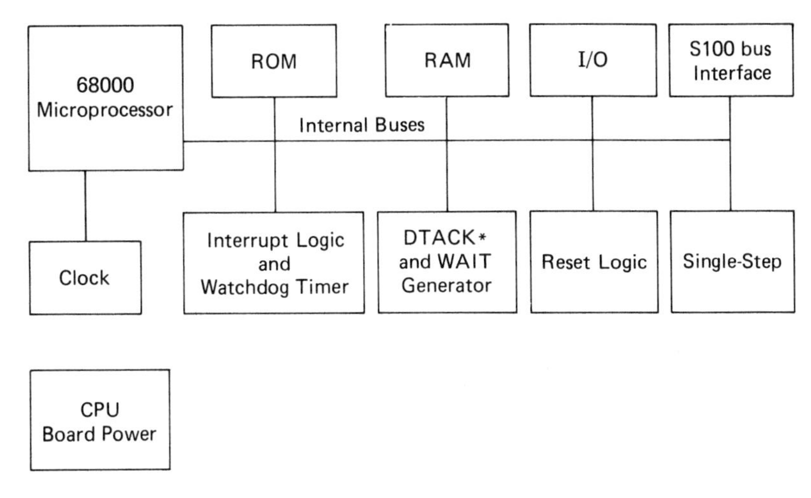

Figure 6.2 Some of the major functions on the CPU board.

with software development: if you develop and write programs in modules, then getting the complete program to work is simple. On the other hand, if you write the whole program at once, then you need sophisticated debugging tools to straighten out a host of interacting program bugs.

The cornerstone is being able to accomplish module testing with an oscilloscope is the ability to freerun the processor. To freerun a processor, you disable the memory-to-CPU feedback loop and provide a no-operation (NOP) instruction every time the microprocessor reads an instruction. The result is that the processor continually cycles through its address range over and over, reading and executing a NOP instruction. Because the instructions are all the same and are repeating, you can easily synchronize an oscilloscope to look at all the processor's address and control lines.

The circuit that implements this freerunning mode is called the "kernel." The kernel is the absolute minimum circuit: if it will not run, then the system cannot function. Put another way, if you have the microprocessor kernel running, then your system has a chance of working. *You can use the free-running kernel as a means of testing all the additional circuits you design and build on your CPU board.*

6.4 SUMMARY

The 68000 project is considerably larger than the clock example in the last chapter, and yet it can be successfully completed by using a plan. It all begins with customer needs; after you understand what is wanted, you can define the project and set specific objectives. The objectives are the product specifications, and they will influence your activity throughout the whole project.

How you actually reach your objectives by building a board that meets specifications depends on your strategy. Rather than following the previous strategy of a complete paper design followed by the building and testing, you might consider an incremental design/build/test strategy that takes smaller steps. By doing this, not only do you avoid being overwhelmed by a huge project, you also develop an easy-to-test system. If you keep module testing in mind as you design, the development of the whole system will be much easier.

Part of designing for testability involves the idea of freerunning the processor so that it constantly cycles over its address range. This allows you to use your oscilloscope to look at the various parts of the system as you refine your design. The minimum circuit that implements the freerunning is the kernel.

Your project plan of action is along the same lines as the clock in Chapter 5: write a do-it list of the tasks that you should complete and a rough schedule to make sure you have time to finish. The plan presented in this chapter has the detail you want for the project; less detail would be too little help in staying on target, and more detail would be unrealistic without experience. As the job progresses, you can keep a daily or weekly list of current tasks you need to follow.

Make a point of checking your schedule of tasks on a daily basis. If you find a particular module extremely difficult, do another one instead if possible; as you do more of the board, your understanding will improve, and you can come back to the troublesome module with a fresh view of how it should work. When you work your schedule, consider this: *a project gets late one day at a time.*

EXERCISES

Note: This sequence of exercises is intended to give direction to the product design shown in the Appendix C Technical Manual. If you intend to build the 68000, write your answers to these exercises in your lab notebook for ready reference during the design and construction.

1. You were asked recently to design a 68000 system and build a prototype model on a solderless breadboard. The customer wants to see if your prototype will be a feasible approach to logging data and performing statistical analysis. Write an outline of what you think the 68000 board might be able to do for him. Write a list of questions you have about uncertain aspects of the project.
2. Suppose you met with the customer and now have a clear idea of the project. Write the project definition as you understand it.
3. List the software requirements (objectives) of the project. Assume that you have the TUTOR EPROM set available to use in your prototype.
4. List the hardware requirements of the project. Assume that you do *not* have a requirement to interface to the IEEE Std-696 bus.
5. List the testing and maintenance requirements. Assume that you have only a logic probe, a multimeter, and a dual-trace oscilloscope.
6. Describe the strategy you will follow to meet your objectives.
7. Write out your plan of action. Include contingency plans so that your project does not get held up by temporary parts shortages or design difficulties.

8. Assume that your whole project must be completed within 12 weeks; at the end of the project, you will make a technical presentation to the customer and deliver your prototype with a complete technical manual. Sketch a schedule showing your plan of work for each of the next 12 weeks.

FURTHER READING

LOCK, DENNIS. *Project Management.* Toronto, Canada: Coles Publishing Company, Ltd., 1980.

RAY, MARTYN S. *Elements of Engineering Design.* London: Prentice-Hall International, UK, Ltd., 1985. (TA 174.R37)

WILCOX, ALAN D. "Bringing Up the 68000—A First Step." *Doctor Dobb's Journal* 11(1): Jan 1986, 60–74.

APPENDIX A

Standards for Schematic Diagrams

A.1 PAPER

Use the same size paper for all the schematics related to the documentation for a single project. Use either

"A" size, 8 ½ × 11 inches, or

"B" size, 11 × 17 inches.

The paper should have zonal coordinates along the borders so that each sector of the drawing can be located in much the same manner as on a roadmap. For example, you might refer to location "A1" in the lower right of the sheet or perhaps location "B2," which is up and to the left of A1.

The paper should be lightly lined with 10 × 10 squares to the inch to assist in drawing. It should not be necessary to use anything more than a straight-edge and a logic-symbol template to do an acceptable drawing.

Put a title block in the lower-right corner of the drawing. Minimum information in the block should be:

Title of the circuit,

Drawing and sheet number,

Revision level of drawing,

Name of draftsman or designer and date.

Use a pencil for all drawings. A dry cleaning pad is helpful to keep the drawing from smudging.

A.2 DIGITAL-LOGIC DRAWING CONVENTIONS

A.2.1

Label all integrated circuits and LSI devices with "U" numbers written inside the symbol for the device. For example, write U2 inside the symbol for one of your logic gates. If there are multiple gates within a single IC, append an a, b, c, etc. to the U number. Put the part number of the device above the symbol for quick part identification. Write the circuit pin numbers outside the symbol. Example:

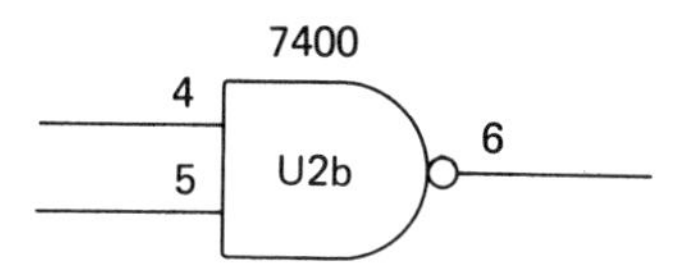

A.2.2

Use mixed-logic symbols to represent the logical functions in the circuit. A NAND gate is identical to an INVERT-OR, and a NOR gate is the same as an INVERT-AND. Use the symbol that indicates how the circuit is intended to perform. Examples:

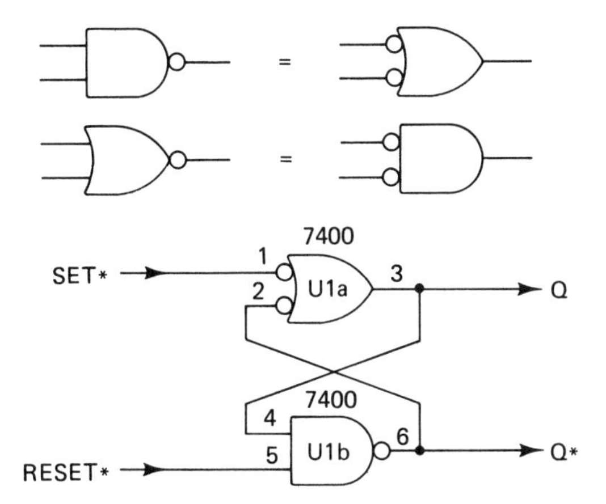

A.2.3

Draw all schematics with the flow of data running left to right or top to bottom. Indicate circuit inputs on left or top and outputs on right or bottom. Show the direction of signals with arrows.

A.2.4

Put function labels *inside* the symbol for logic devices. For example, if a device has inputs A, B, and ENABLE, then show those labels inside the symbol. Example:

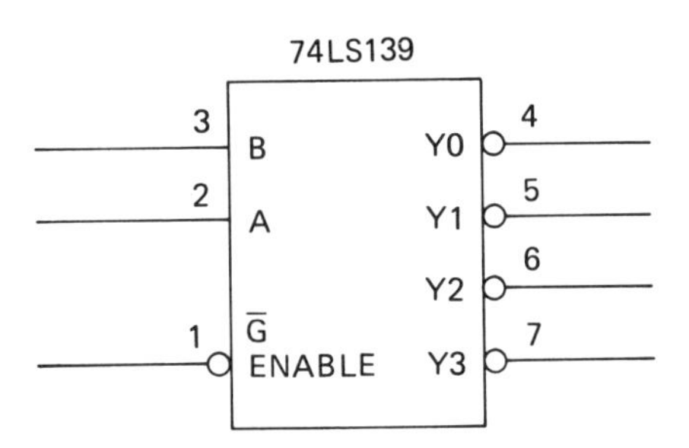

A.2.5

Put bubbles at inputs and outputs as indicated on manufacturer data sheets. For example, if an output is understood to be low when asserted, then a bubble will be indicated at the output; see 74LS139 outputs in A.2.4.

If the manufacturer indicates a function label with an overbar (negative logic), indicate the asserted-low nature of the input or output line by appending a "star." Example:

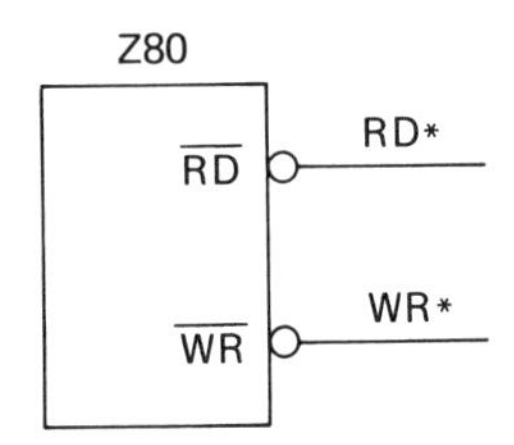

A.2.6

Provide page-to-page connections by using the zonal coordinates of the source signal and the destination. Indicate the name of the signal line on both ends and show the "to" and "from" coordinates. The example below shows a signal named RAS* leaving sheet 1 at location A1 and going to sheet 3 location B2. Be sure to draw arrows indicating the direction of the signals.

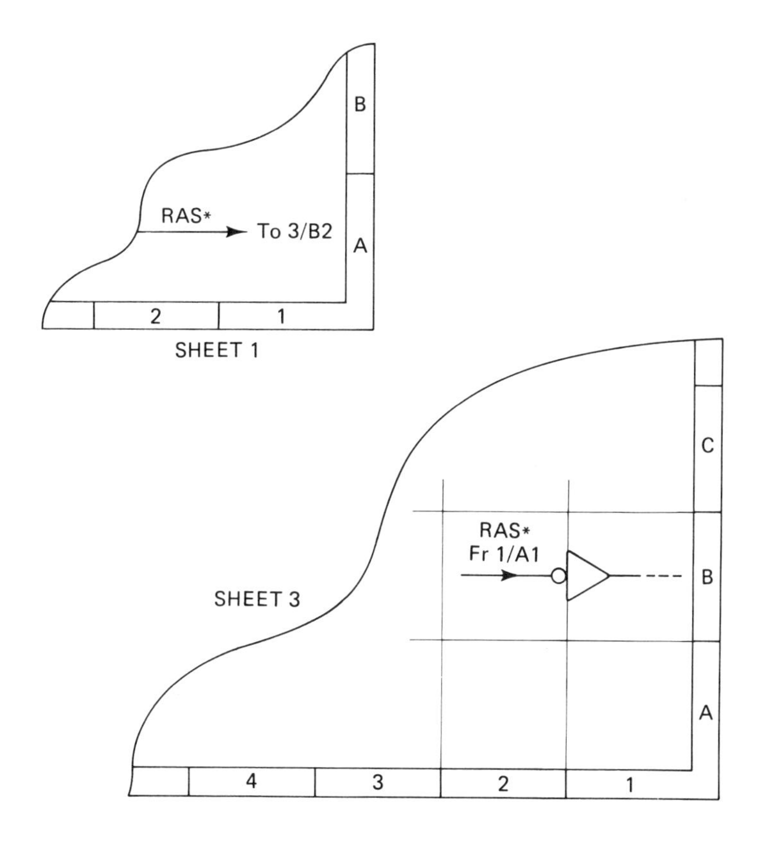

A.2.7

Draw any unused spare gates at the corner of the schematic. Tie all TTL inputs high with a 1K resistor. Example:

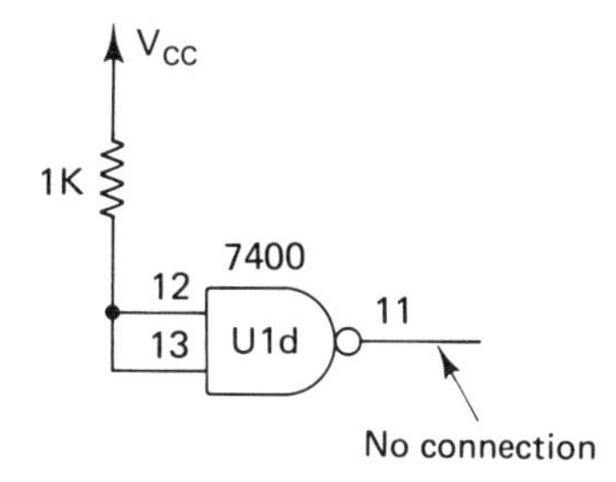

A.2.8

Provide a power table with the schematic diagram to account for power and grounds to all devices. Example:

Part	*Type*	*Ground*	*+Vcc*
U1	7400	7	14
U2	7404	7	14
U3	74LS139	8	16

A.2.9

Draw crossing wires as shown below:

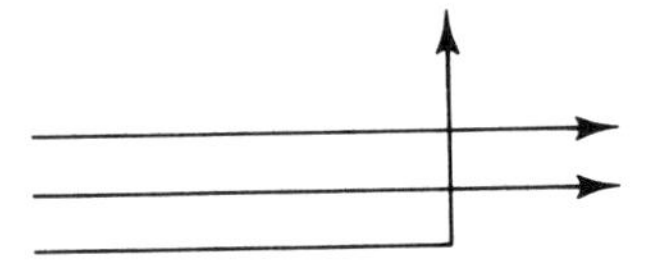

When several wires connect to a common line, connect with a dot and draw the connections slightly offset so they are not mistaken for crossing wires. Example:

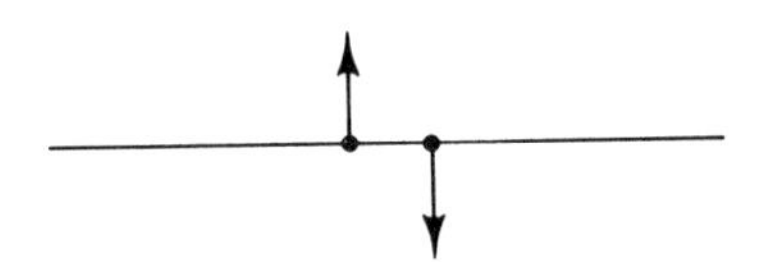

A.2.10

Draw multiwire buses as a heavy line with a slash across it and a number to indicate the wire count. Show the signal name for any wires breaking from the bus. Example:

A.2.11

Draw grounds with the symbol: ⏚

Show connection to +5 V supply:

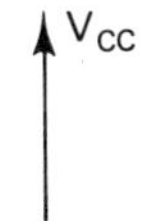

Indicate pullup resistors either as shown in A.2.7 or use the symbol: △

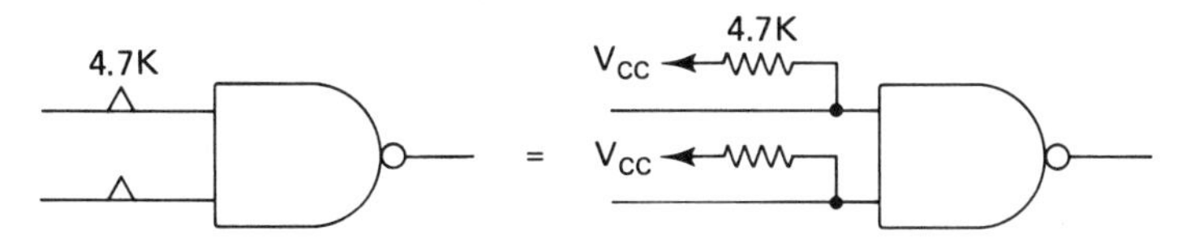

A.3 TYPICAL DRAWING

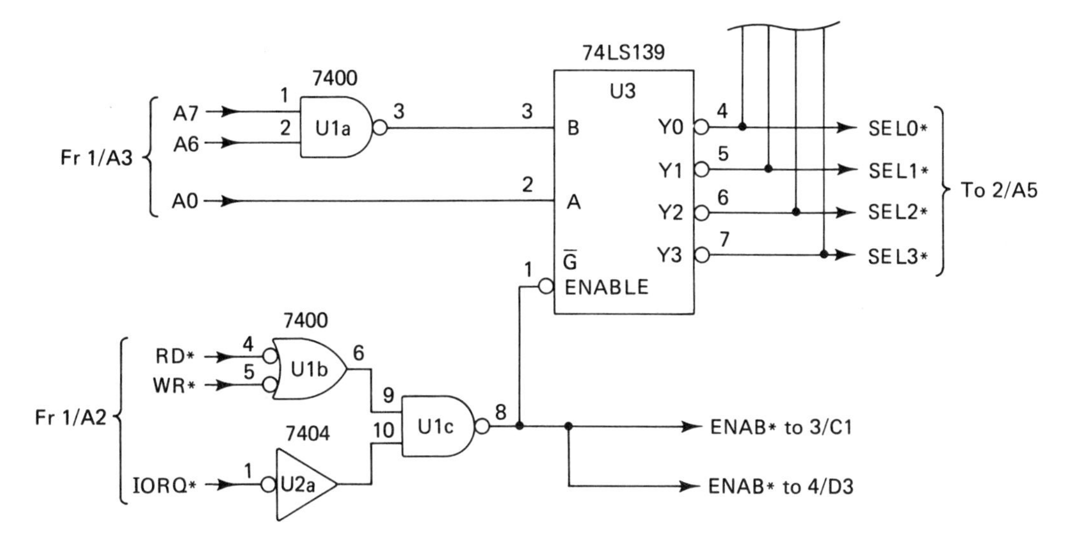

APPENDIX B

Temperature Monitor Technical Manual: Hot-Spot Oil Company

Alan D. Wilcox

March 1987

INTRODUCTION

The purpose of the temperature monitor is to measure and record outside air temperature every hour each day and then calculate heating degree-days over a 24-hour period.

During the heating season it is necessary to know how cold the weather has been each day so that timely oil deliveries can be made. Deliveries made too frequently result in extra driver time and truck mileage as well as extra administrative billing costs. On the other hand, a late delivery can result in delivery driver overtime and in possible loss of customers. If the heating degree-days can be totaled since the last oil delivery, then it is possible to estimate when the next delivery should be made. The temperature monitor provides the basic information needed to make this estimate.

The temperature monitor is a small self-contained unit that can be placed on a desk or other convenient location in the office. A single wire with a temperature sensor on the end is put outside and connected to the monitor. The monitor runs on standard house current, but has an internal battery to retain data in case of power failure. The temperature and degree-days are displayed on the front panel; the time and latest 24-hour cumulative degree-days are printed each hour on a paper tape in the unit.

After setting up the monitor, the paper tape can be checked each morning to see how many heating degree-days were needed during the last 24 hours. This reading can be added to the total degree-days accumulated for each customer since fillup. When the customer total reaches a certain threshold, say 1100 degree-days, it is time to make an oil delivery to that customer.

INSTALLATION

As shown in Figure B-1, place the temperature monitor in a convenient location where it will be used during normal daily operation. Plug the power cord into a wall socket.

Install the temperature sensor outside in a shaded spot and run the wire into the building to the monitor. Attach it to the connector on the rear of the unit.

Press the button marked "test" on the front panel. The monitor will check itself and the sensor you just connected. It will display "OK" in several seconds at the end of the test and you can begin normal operation. If you do not get the "OK" display, refer to the Troubleshooting section in this manual.

OPERATION

To begin operation, push the "start" button on the front panel. You will see a display of the present outside temperature. No data will be recorded on the paper tape until after an hour has elapsed. The correct present outside temperature will be displayed continuously.

To set the time, push the "set time" button on the front panel. Press four numbers on the keyboard for the correct time. All time is maintained in 24-hour format; that is, 1 PM is 1300, 11 PM is 2300, etc. Example entries:

Time is →	7:00 A.M.	Press →	0700
	11:35 A.M.		1135
	8:40 P.M.		2040

To set the time at which the data logging is done, press the "start log" button on the front panel. This should be done on the hour if you want your time and degree-day prints recorded on the hour. If you prefer the prints recorded hourly on the half-hour, then press the "start log" button when the time is half-past the hour.

Note that the first day's reading of degree-days will not be correct until 24 hours have past. After that, the display always shows the correct heating degree-days regardless of when the data log is printed. Every hourly printout will correctly represent the degree-days during the most recent 24 hours.

CIRCUIT DESCRIPTION

The temperature monitor measures the current temperature and continuously displays the temperature and the latest 24-hour heating degree-day summary. Every hour the time and degree-days are printed on a paper tape in the unit. The monitor uses a microcomputer to perform the control of the system and do the various calculations.

The temperature monitor is composed of a number of subsystems as shown in Figure B-2. The temperature sensor is connected to the analog-to-digital (A/D) converter; the output of the A/D converter is interfaced to the microcomputer itself. The operation of the keypad, display, and paper-tape printer are all controlled by the microcomputer.

The temperature sensor, the analog circuits, and the digital interface control for the A/D are shown in the schematic diagram of the unit.

SOFTWARE DESCRIPTION

The temperature monitor software is contained in permanent memory in the microcomputer section. The programs handle all the operations necessary to initialize the unit, to test for proper performance, to set the time, to set the logging time, to read and display temperature, to calculate and display heating degree-days, and to print temperature and degree-days.

The software functions are shown in the structure chart in Figure B-3. The temperature monitor software is divided into four main modules: SELF-TEST, SET-TIME, SET-LOG-TIME, and RUN. When the computer is first turned on, the SELF-TEST code should be executed to verify the system operation and to initialize data memory. The SET-TIME module is used to set the system internal time clock. The SET-LOG-TIME module is used to set when each hour the temperature data is printed on the paper tape. The last module, RUN, is used to control the normal system operation.

When the RUN module is executed, the outside temperature is read once each second, averaged over the last 16 seconds, and the average saved in memory. Each hour all these averages are averaged again and used to calculate the hourly degree-days using the formula "degree-days = 65 − average hourly temperature." The most recent hourly degree-days are saved in memory and averaged together to display and record on the paper tape.

TROUBLESHOOTING

The temperature monitor has an internal self-test feature that will indicate the possible cause of various problems. To use, press the "test" button on the front panel; the display will either indicate "OK" if the system is ready or indicate an error number if there is a problem. Some of the symptoms and their causes are listed in the chart below:

Condition	*Possible Cause/Defect*
Nothing happens at turn-on	Not plugged in Fuse blown Power supply
Fuse blows when plugged in	Short in power supply
Display shows random digits at turn-on	Microprocessor Main memory

Condition	Possible Cause/Defect
When "test" button pressed, the display shows . . .	
blank or random digits	Microprocessor Main memory
01 (Memory error)	Data memory
02 (No temperature)	Temperature sensor Analog amplifier A/D converter
03 (Printer)	Printer out of paper

REFERENCES

This manual contains the technical information on the temperature monitor analog and interface circuits. The circuit diagram and parts layouts for both of these subsystems are provided in the appendix to this manual. Information on the microcomputer subsystem and the computer program listings are not provided; they may be obtained from the factory on special order.

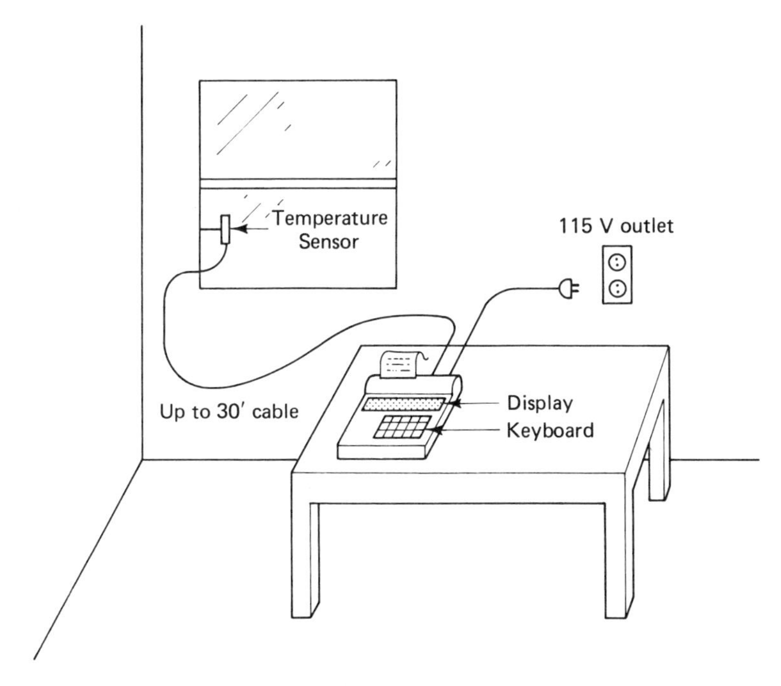

Figure B-1 Temperature monitor setup for normal operation.

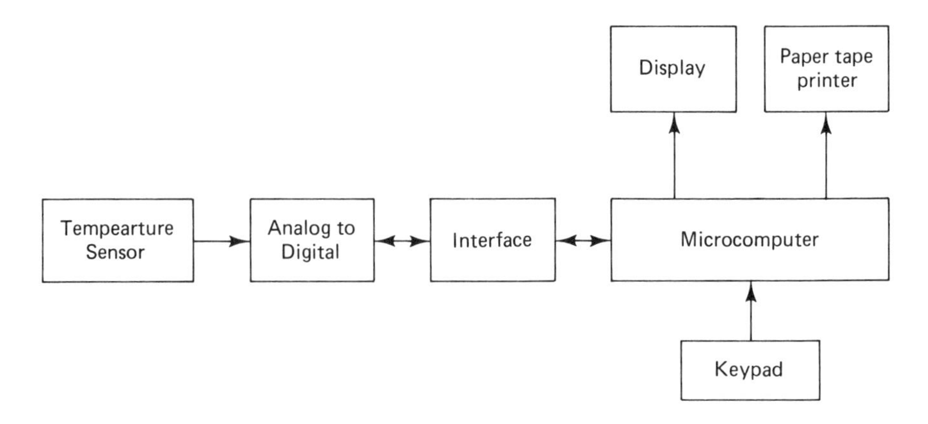

Figure B-2 Block diagram of temperature monitor.

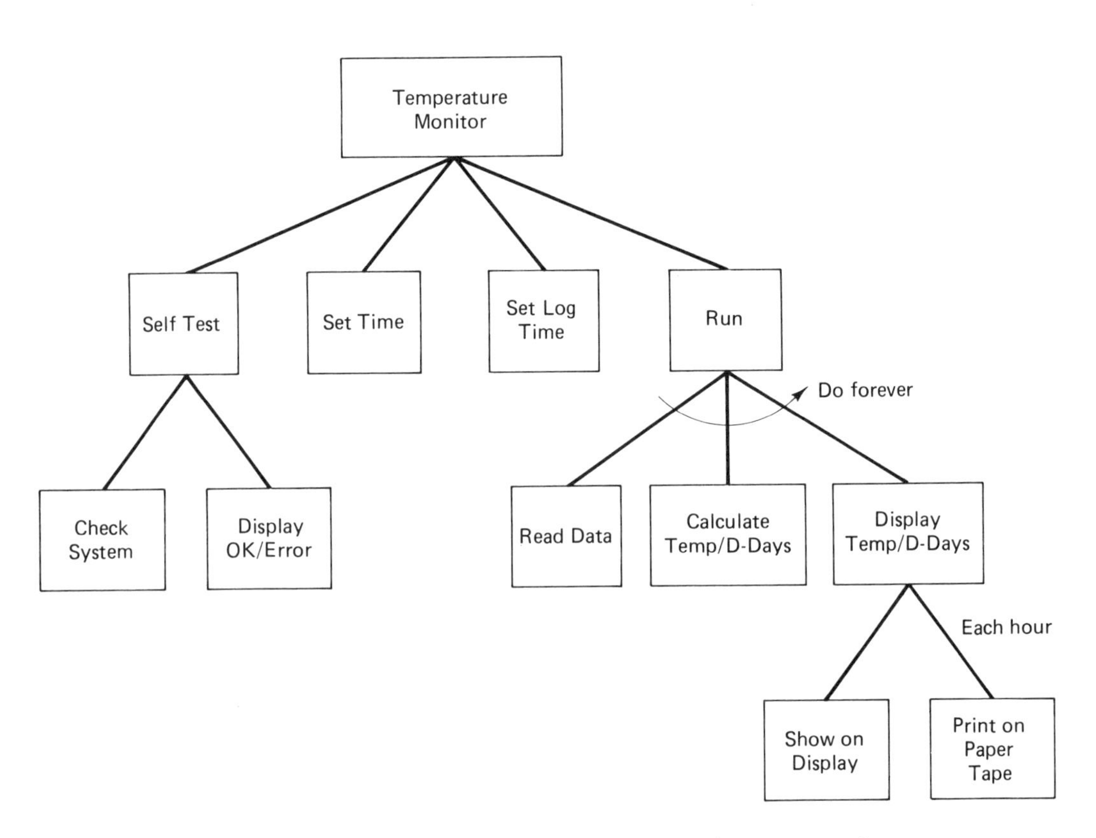

Figure B-3 Temperature monitor system-software structure chart.

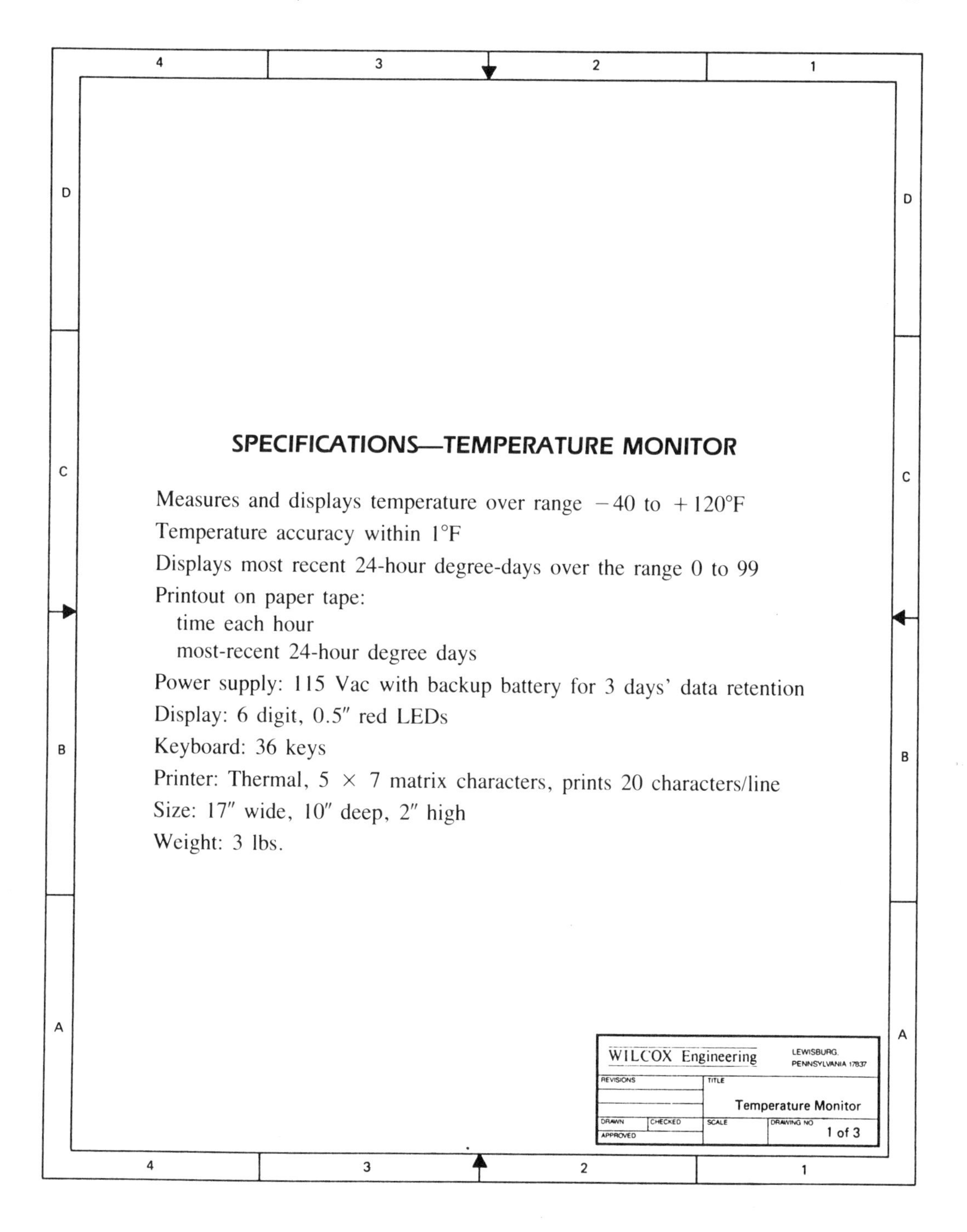

SPECIFICATIONS—TEMPERATURE MONITOR

Measures and displays temperature over range −40 to +120°F
Temperature accuracy within 1°F
Displays most recent 24-hour degree-days over the range 0 to 99
Printout on paper tape:
  time each hour
  most-recent 24-hour degree days
Power supply: 115 Vac with backup battery for 3 days' data retention
Display: 6 digit, 0.5″ red LEDs
Keyboard: 36 keys
Printer: Thermal, 5 × 7 matrix characters, prints 20 characters/line
Size: 17″ wide, 10″ deep, 2″ high
Weight: 3 lbs.

WILCOX Engineering LEWISBURG, PENNSYLVANIA 17837
REVISIONS
TITLE Temperature Monitor
DRAWN CHECKED SCALE DRAWING NO 1 of 3
APPROVED

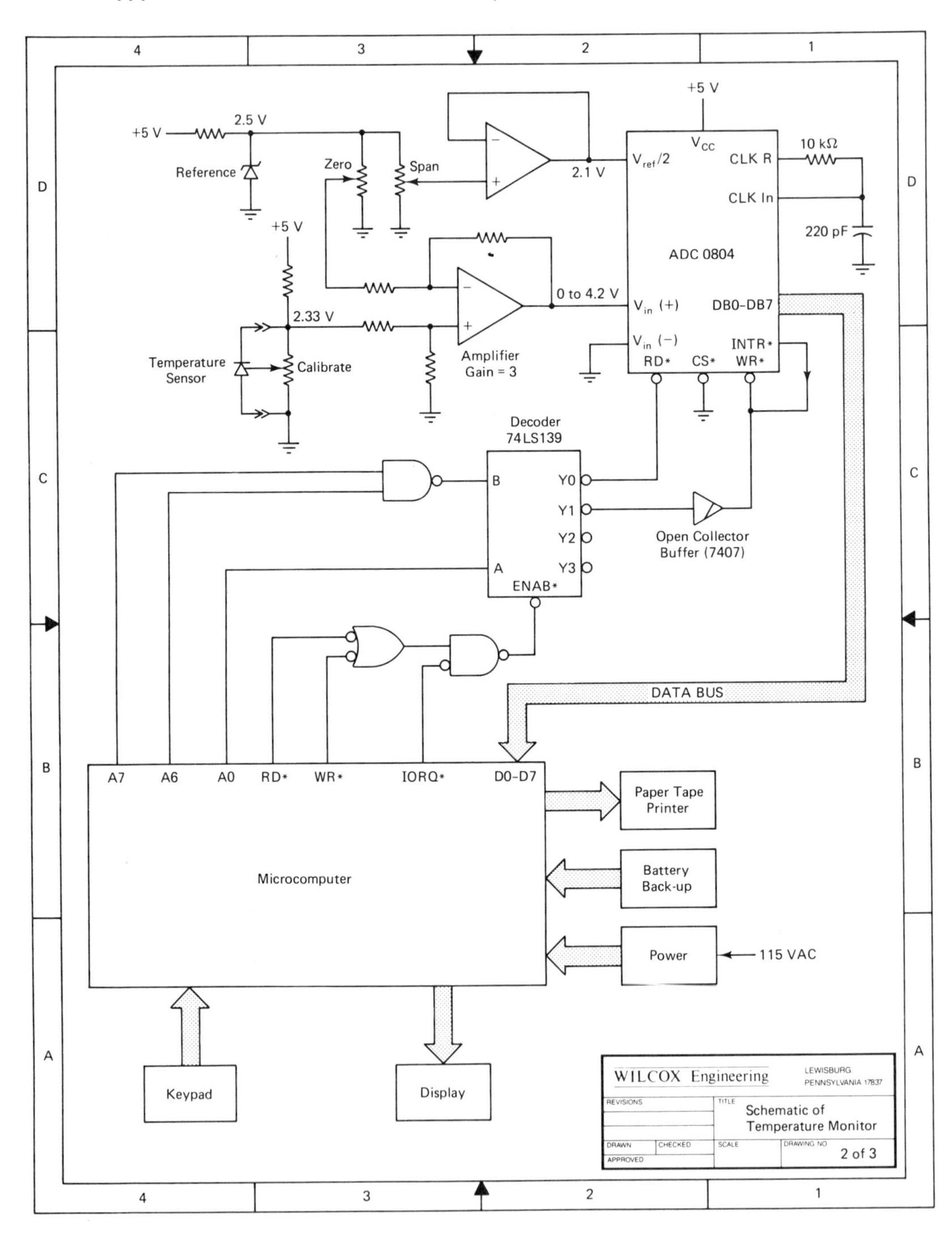

+5 V
2.5 V
Reference
Zero
Span
2.1 V
+5 V
V_{CC}
$V_{ref}/2$
CLK R
10 kΩ
CLK In
220 pF
ADC 0804
+5 V
0 to 4.2 V
2.33 V
V_{in} (+)
DB0–DB7
V_{in} (−)
INTR*
RD*
CS*
WR*
Temperature Sensor
Calibrate
Amplifier Gain = 3
Decoder 74LS139
B
Y0
Y1
Y2
Y3
A
ENAB*
Open Collector Buffer (7407)
DATA BUS
A7
A6
A0
RD*
WR*
IORQ*
D0–D7
Microcomputer
Paper Tape Printer
Battery Back-up
Power
115 VAC
Keypad
Display
WILCOX Engineering
LEWISBURG PENNSYLVANIA 17837
REVISIONS
TITLE
Schematic of Temperature Monitor
DRAWN
CHECKED
SCALE
DRAWING NO
APPROVED
2 of 3

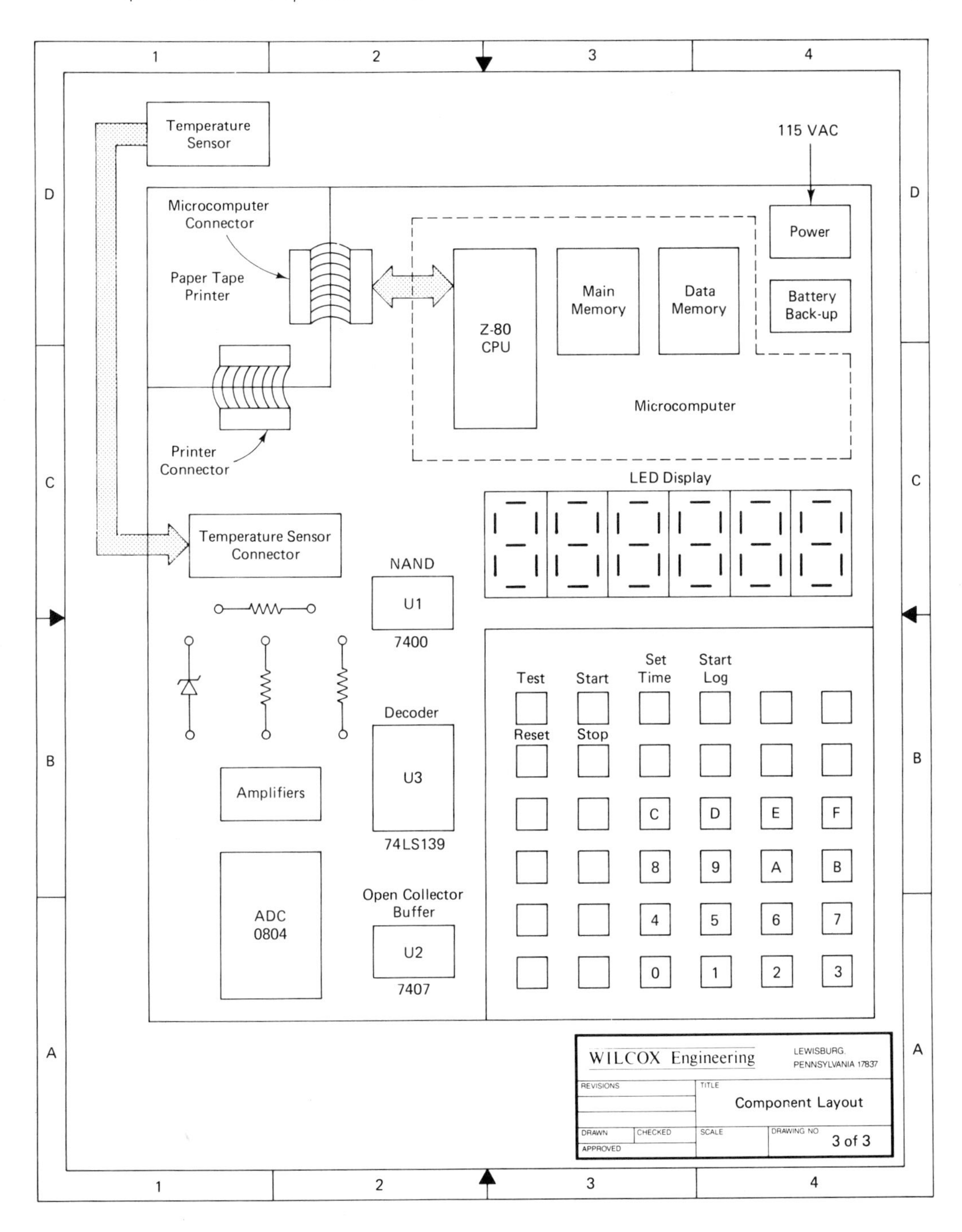
1
2
3
4
D
C
B
A
Temperature Sensor
115 VAC
Microcomputer Connector
Paper Tape Printer
Power
Battery Back-up
Main Memory
Data Memory
Z-80 CPU
Microcomputer
Printer Connector
LED Display
Temperature Sensor Connector
NAND
U1
7400
Test
Start
Set Time
Start Log
Reset
Stop
Decoder
U3
74LS139
Amplifiers
C
D
E
F
8
9
A
B
4
5
6
7
0
1
2
3
ADC 0804
Open Collector Buffer
U2
7407
WILCOX Engineering
LEWISBURG, PENNSYLVANIA 17837
REVISIONS
TITLE
Component Layout
DRAWN
CHECKED
SCALE
DRAWING NO
3 of 3
APPROVED

APPENDIX C

68000-Based CPU Board Technical Manual

Kanwalinder Singh

December 16, 1986

TABLE OF CONTENTS

CHAPTER 1 INTRODUCTION

The purpose of the KS68 is to provide the user with an introducti on to systems based on the Motorola 68000 family of microprocessors. Located on a single protoboard, a complete microprocessor system is provided, including an MC68000 16- bit microprocessor, memory, and a serial communication I/O, besides on-board troubleshooting aids. The user has to connect an RS-232C compatible "dumb" terminal and the power supplies to have a functional system.

For easy use, the board has a resident firmware package that provides a self-contained programming and operating environment. The firmware provides the user with monitor/debug, assembly/disassembly, program entry, and I/O control functions. Utilizing the capabilities provided by the system, the user can investigate and learn the computing power and the architectural features of the 68000. Being a working example of the 68000 external bus structure and interface to memory and peripherals, the KS68 also provides the user with a reference for similar design and/or expansion.

The KS68 features include:

a. 4 MHz (MC68000) 16-bit MPU.
b. Clock speed—8 MHz (max.)
c. 4K bytes of static RAM (6116) arranged as 2K × 16.
d. 16K bytes firmware EPROM (27128) arranged as 8K × 16.
e. 4K bytes of user EPROM (2716) arranged as 2K × 16.
f. One serial, RS-232C compatible, baud rate selectable communication port provided for a terminal.
g. Self-contained operating firmware that provides monitor, debug, and disassembly/assembly functions.
h. RESET and SINGLE-STEP function switches.

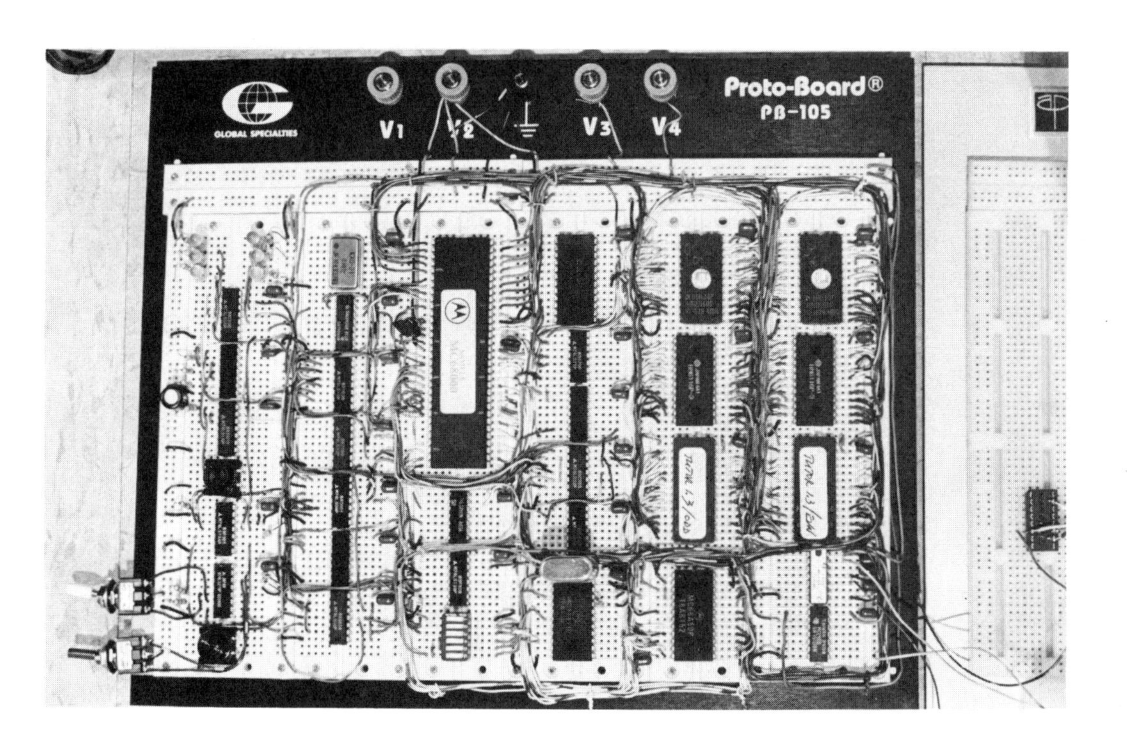

SPECIFICATIONS

Microprocessor	MC68000
Input/output	One serial, RS-232C compatible, baud rate selectable (300, 1200, and 9600 baud) communication port provided for a terminal.
System clock	4 MHz
Memory	4 Kb RAM arranged as 2K × 16, accessible on a byte or word basis. 4Kb EPROM arranged as 2K × 16, accessible on byte or word basis.
Software	16 Kb TUTOR provides monitor, debug, assembly/disassembly, program entry, and I/O control functions.
Waits	Jumper selectable 0 to 7 waits on all memory operations.
Reset	Boot-up circuitry selects the TUTOR EPROM during the first eight bus cycles, after reset or on power-up, providing the 68000 with the stack pointer and the program counter. During normal operation, RAM is available in lower memory.
Displays	Individual LEDs indicate: 68000 halted condition, freerun, DTACK* status, RAM chip enable, and TUTOR EPROM chip enable.
Control	Switches are provided to single-step the 68000 by delaying DTACK*; all bus signals remain valid and can be easily checked during troubleshooting.
Power requirements (typical)	+5.0 V/750 mA, +12 V/50 mA, −12 V/50 mA.
Operating temp.	0 to 50°C.
Board dimensions	9.2″ × 11.4″ × 1.5″ (L × W × H).

CHAPTER 2 INSTALLATION AND POWER-UP INSTRUCTIONS

2.1 Preparing the Board for Use

Figure 2.1 shows the layout of the KS68. Board preparation involves the following steps:

a. Make the power connections by connecting V2(+5V), V3(+12V), V4(−12V), and GND(GND).
b. Set the single-step switch (SW6) to RUN.
c. Check that jumpers J1 and J2 are in place.
d. Set switches SW2, SW3, and SW4 to OFF.
e. Select communication baud rate using SW7(9600), SW8(1200), or SW9(300).

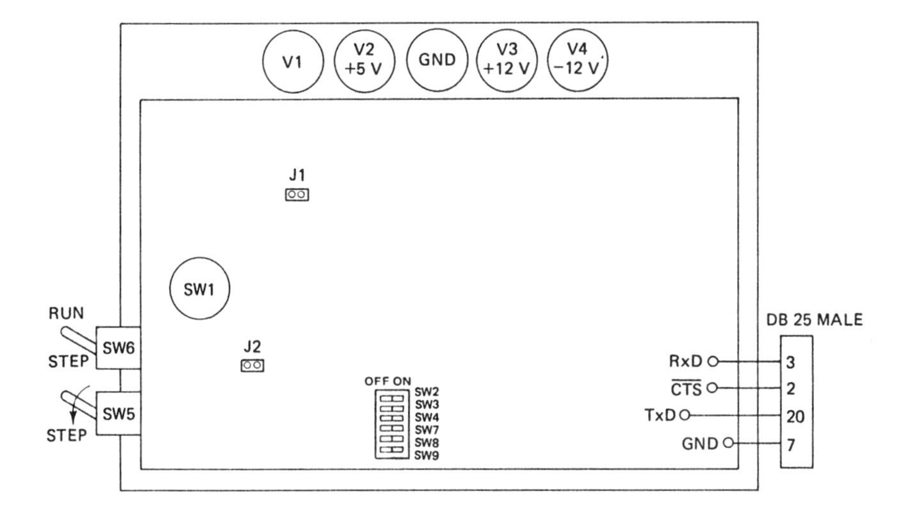

Figure 2.1 KS68 board layout.

2.2 System Hook-Up Instructions

Connect the board to the terminal via the connector provided as shown in Figure 2.2. Check that the terminal and the board are set to the same baud rate.

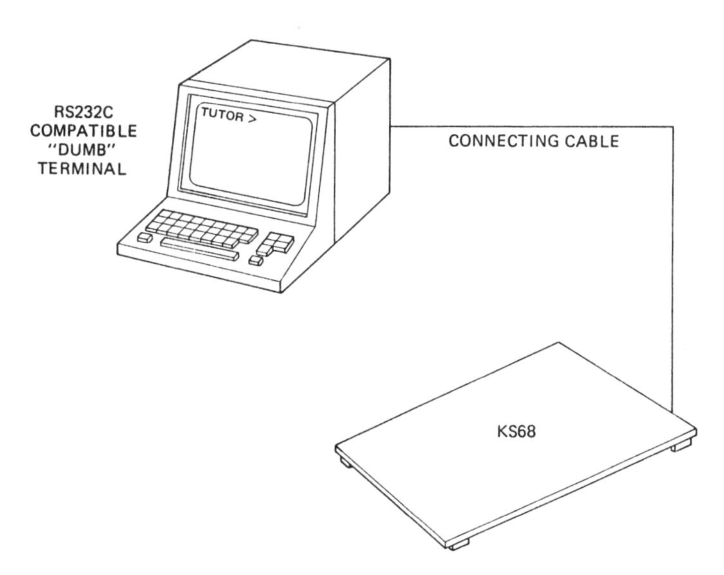

Figure 2.2 System hookup.

2.3 System Turn-On and Initial Operation

a. Be sure all voltages are connected to the board prior to power-up.
b. Turn the power ON.

On power-up, the system should initialize itself and print on the terminal:

TUTOR 1.3>

It is now ready for operation under the control of the firmware. If this response does not appear, perform the following system checks:

a. Press the reset switch (SW1) to ensure that the board has been initialized properly.
b. Check that the terminal and the board are set for the same baud rates.

If the baud rates are set properly and terminal is still not responding properly, the board may require some detailed system checks. Refer to Chapter 5 for details.

CHAPTER 3 OPERATING INSTRUCTIONS

3.1 System Operation

After system initialization or return of control to TUTOR, the terminal will print

TUTOR 1.3 >

and wait for a response.

The user can call any of the commands supported by the firmware. (Refer to Chapters 3, 4, and 5 of MC68000 Educational Computer Board User's Manual for detailed information.) A standard input routine controls the system while the user types a line of input. Command processing begins only after a line has been entered, followed by a carriage return. It may be noted that:

a. The user memory is located at addresses \$000900–\$000FFF. When first learning the system, the user should restrict his activities to this area of the memory map.
b. As the board does not have a bus error control circuit, if a command causes the system to access an unused address (i.e., no memory or peripheral devices are located at that address), the system will just "hang" and do nothing. Press the RESET switch to recover from such a situation.

CHAPTER 4 HARDWARE DESCRIPTION

The functional block diagram of the KS68 is shown in Figure 4.1. The various modules that form the complete system are described below. Before proceeding further, the user is

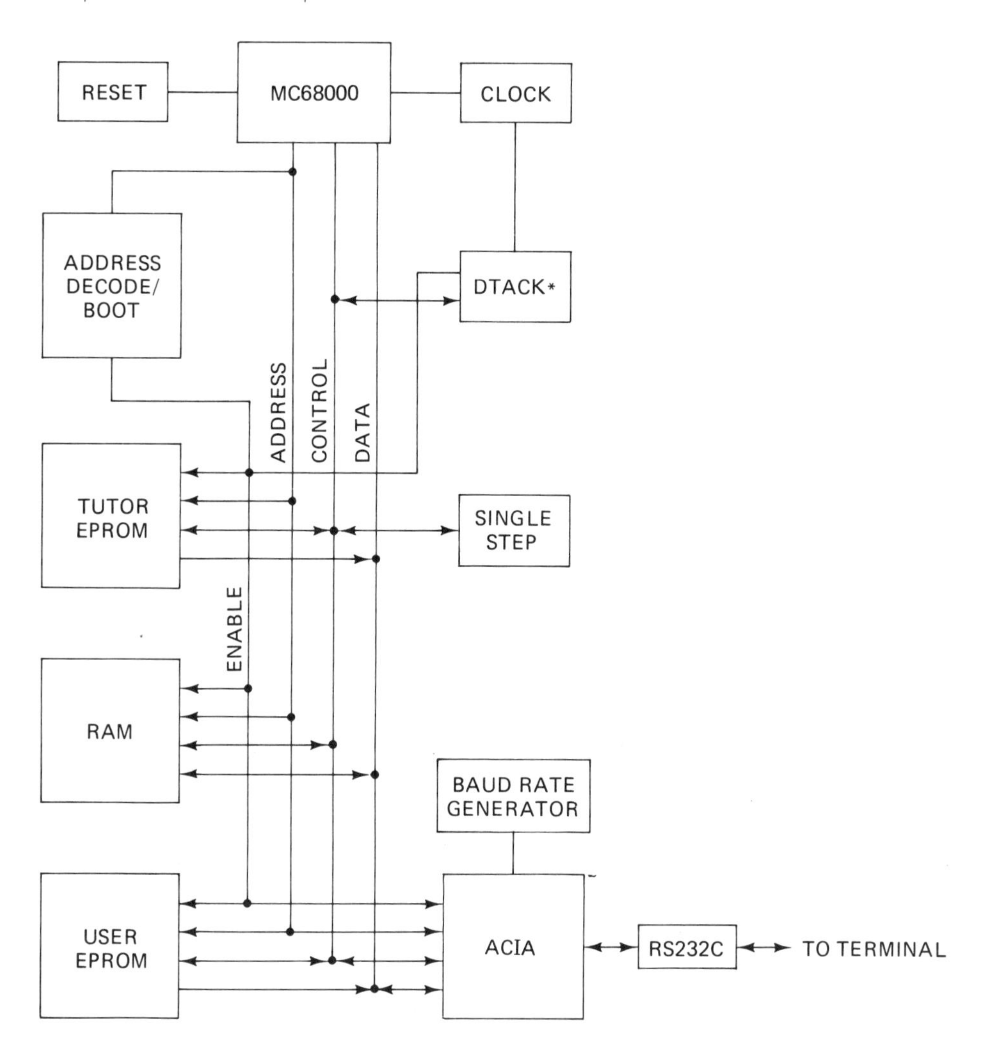

Figure 4.1 Functional block diagram.

advised to go through the 68000 signal and bus operation description given in Section 4 of MC68000 Information Manual.

4.1 Test Module

Reference: Drawing number S-02.

This module monitors the state of some important signals on the board, as shown in Table 4.1.

TABLE 4.1 TEST MODULE CONFIGURATION.

LED #	*Signal name*	*Signal description*	*State when LED will glow*
1.	HALT*	Halt line	Low
2.	A20	Address bit 20	High
3.	DTACK*	Data transfer acknowledge	Low
4.	CETR*	TUTOR EPROM chip enable	Low
5.	CER*	RAM chip enable	Low

4.2 Clock Module

Reference: Drawing number S-03.

This module provides the 68000 clock (CLK68) and an inverted clock (SYSCLK) with a typical skew of 0.5 ns. Both the outputs sink 32 mA (2 × 16 mA) and source 1.6 mA (2 × 0.8 mA).

4.3 Reset Module

Reference: Drawing number S-04.

This module keeps the RESET* and HALT* asserted for about 300 ms, both on power-up and on switch reset. Pressing the RESET switch (SW1) causes all processes to terminate, resets the MC68000 processor and restarts the TUTOR firmware. Pressing this switch should be the appropriate action if all else fails.

4.4 Freerun Module

Reference: Drawing number S-05.

The MC68000 can be made to freerun at any stage by plugging in the 24-pin headers (provided with the board) into the TEST EPROM sockets, by disabling J1 and by putting switches SW2 and SW3 to ON. A steadily blinking LED2 will indicate proper freerun.

4.5 DTACK* Module

Reference: Drawing number S-06.

This module returns DTACK* to the 68000 every time it performs a data operation on the board's memory devices. The module is set up to return DTACK* before the falling edge of state 4 (0 waits) but it can be configured to provide up to 7 waits using jumper J2.

4.6 Address Decode/Boot Module

Reference: Drawing numbers S-07 and S-09.

The memory map of the KS68 is shown in Table 4.2. The RAM is addressed at the bottom of the map ($000000–$000FFF) except during the initialization sequence when the

TABLE 4.2 MEMORY MAP.

Function			*Address*
System memory	Exception vector table	EPROM	$000000–$000007(*)
		RAM	$000000–$0003FF
	TUTOR scratchpad	RAM	$000400–$0008FF
User memory		RAM	$000900–$000FFF
Not used(**)			$001000–$007FFF
TUTOR firmware		EPROM	$008000–$00BFFF
User memory		EPROM	$00C000–$00CFFF
Not used(**)			$00D000–$00FFFF
I/O		ACIA	$010040&$010042
	Redundant mapping		↓ $01FFFF
Not used(**)			$020000–$07FFFF
Not used(***)			$080000–$FFFFFF

* Only during the initialization sequence.
** Decoded on the board—available for future expansion.
***Not decoded on the board.

TUTOR EPROM is decoded at locations $000000–$000007. The RAM is divided into two areas: addresses $000000–$0008FF are the system area reserved for use by the system firmware, and addresses $000900–$000FFF are the user area. Within the system area, addresses $000000–$0003FF are used for the MC68000 exception vector table. The remaining 1,280 bytes (addresses $000400–$0008FF) are used as scratchpad memory by the TUTOR firmware for data buffers, pointers, temporary storage, etc.

The firmware EPROM is located at addresses $008000–$00BFFF. Moreover, a user EPROM is provided at addresses $00C000–$00CFFF to hold user-generated code.

The 6850 ACIA is mapped at addresses $010040 and $010042. An additional ACIA for a host can be put at addresses $010041 and $010043. The ACIA address decode is redundant within the page $01XXXX. The ACIA can be accessed any time address line A6 = 1 within this page.

Additional areas, as shown in Table 4.2, are decoded on the board for future expansion.

4.7 RAM Module

Reference: Drawing number S-11.

This consists of two 6116 static RAMs arranged as 2K × 16. These can be accessed either on a byte or word basis, using asynchronous data transfer.

4.8 TUTOR EPROM Module

Reference: Drawing number S-12.

The system firmware (TUTOR) is stored in two 27128 EPROMs, half of which are empty. The system EPROM can be read on a byte or word basis. Attempting to write into the EPROM will result in a bus timeout error.

4.9 User EPROM Module

Reference: Drawing number S-08.

The user-generated code can be stored in two 2716 EPROMs provided on the board. These EPROMs are also used for on-board troubleshooting. The user EPROMs can also be read on a byte or word basis.

4.10 Single-Step Module

Reference: Drawing number S-10.

The MC68000 can be stepped through the program space one bus cycle at a time by putting the single-step switch SW6 in the STEP position. The 68000 executes a bus cycle every time SW5 is toggled. This circuit is very useful for on-board troubleshooting.

4.11 I/O Module

Reference: Drawing number S-13.

A single 6850 ACIA connected to the data bits D08-D15 provides serial communication (with handshake) with an RS-232C compatible terminal. The 68000 accesses the ACIA (which is a synchronous device) using a synchronous type of bus transfer involving VPA*, VMA*, and the E clock (E = 400 kHz). A 14411 baud rate generator provides transmit and receive clocks for the ACIA. A pair of 1488 and 1489 line drivers is used to translate the ACIA voltage levels to RS-232C interface levels.

CHAPTER 5 TROUBLESHOOTING

The KS68 uses the freerun and the single-step technique to check all the modules, as described in Table 5.1. The user may use a single test or a group of tests described below to isolate a malfunctioning module. The particular module can then be diagnosed using the module schematics.

TABLE 5.1 TROUBLESHOOTING CHART.

Module name	*Operator function*	*Expected Results*
Clock module	1. Check CLK68 and SYSCLK.	1. See Fig. T-01.
Reset module	1. Check RESET* vs. VCC and SW1.	1. See Figs. T-02 and T-03.
Freerun module	1. Plug in the 24-pin headers into the TEST EPROM sockets.	
	2. Disable J1.	
	3. Enable SW2 and SW3.	
	4. Check all address and data pins, on freerun, with a logic probe.	4. All address pins should be pulsing, and all data pins should be low.
	5. Check AS* vs. CLK68.	5. See Figs. T-04 and T-05.
DTACK* module	1. Disable SW2.	
	2. Put J2 to 0 waits.	2. See Fig. T-05.
	3. Put J2 to 7 waits.	3. See Fig. T-06.
Decode module	1. Check CSR*, CSTR*, CST* and CSIO*.	1. See Fig. T-07.
Test EPROM module	1. Disable SW3 and enable J1.	
	2. Remove freerun headers.	
	3. Put CSTR* into TEST EPROM chip select (pin number 18).	
	4. Plug in TEST EPROMs with scoop-loop 1.*	
	5. Check various address and data pins on run.	5. See Table A-1 for address and data pins status.
Boot module	1. Check BOOT* vs. RESET.*	1. See Fig. T-08.
	2. Check CSR*, CSTR*, BOOT*, CER* and CETR* vs. RESET*.	2. See Fig. T-09.
	3. Put CETR* into TEST EPROM chip select (pin number 18).	
	4. Plug in TEST EPROMs with scope loop2.*	
	5. Check various address and data pins on run.	5. See Table A-2 for address and data pins status.
Single-step module	1. Put SW6 on STEP.	
	2. Single-step scope-loop2* from power-up.	2. See Table A-2 for address and data pins status.

*See Appendix A.

TABLE 5.1 (*Cont'd.*) TROUBLESHOOTING CHART.

RAM module	1. Plug in TEST EPROMs with scope loop3.*	
	2. Single-step scope-loop3 from power-up.	
	3. Check data being written and read from RAM at step 9 and 12 respectively.	3. The data should be $FEDC.
TUTOR module	1. Put CETR* to TUTOR EPROMs chip select (pin number 20).	
	2. Check TUTOR startup sequence.	2. See Fig. T-10.
I/O module	1. Check E, VPA* and VMA* on TUTOR run.	1. See Fig. T-11.
	2. Check RTCLK and T_XD vs. RESET*.	2. See Fig. T-12.
	3. Check R_XD with a logic probe while pressing the BREAK key on the terminal.	3. R_XD should change state from high to low on key depression.

*See Appendix A.

REFERENCES

MC68000 Educational Computer Board User's Manual, MEX68KECB/D2, 2nd Edition, Tempe, AZ: Motorola Literature Distribution Center, 1982.

MC68000 16-bit Microprocessor Data Manual, Austin, TX: Motorola Semiconductor Products, Inc.

The TTL Data Book, Vol. 2, Dallas, TX; Texas Instruments, Inc., 1985.

WILCOX, ALAN D. *68000 Microcomputer Systems: Designing and Troubleshooting*. Englewood Cliffs, NJ; Prentice-Hall, Inc., 1987.

APPENDIX A SCOPE LOOPS

TABLE A-1 SCOPE LOOP1.

	Even	*Odd*	
000000	00	00	SSP to 000444
000002	04	44	
000004	00	00	PC to 000008
000006	00	08	
000008	4E	F8	JMP.S $000008
00000A	00	08	

Data Lines

D15	14	13	12	11	10	9	8	7	6	5	4	3	2	1	0	
0	1	0	0	1	1	1	0	1	1	1	1	1	0	0	0	000008
0	0	0	0	0	0	0	0	0	0	0	0	1	0	0	0	00000A

Address Lines

A15	14	13	12	11	10	9	8	7	6	5	4	3	2	1	0	
0	0	0	0	0	0	0	0	0	0	0	0	1	0	0	0	000008
0	0	0	0	0	0	0	0	0	0	0	0	1	0	1	0	00000A

TABLE A-2 SCOPE LOOP2.

	Even	*Odd*	
000000	00	00	SSP to 000444
000002	04	44	
000004	00	00	PC to 008008
000006	80	08	
008008	4E	F9	
00800A	00	00	JMP.L $008008
00800C	80	08	

Data Lines

D15	14	13	12	11	10	9	8	7	6	5	4	3	2	1	0	
0	1	0	0	1	1	1	0	1	1	1	1	1	0	0	1	008008
0	0	0	0	0	0	0	0	0	0	0	0	0	0	0	0	00800A
1	0	0	0	0	0	0	0	0	0	0	0	1	0	0	0	00800C

Address Lines

A15	14	13	12	11	10	9	8	7	6	5	4	3	2	1	0	
1	0	0	0	0	0	0	0	0	0	0	0	1	0	0	0	008008
1	0	0	0	0	0	0	0	0	0	0	0	1	0	1	0	00800A
1	0	0	0	0	0	0	0	0	0	0	0	1	1	0	0	00800C

TABLE A-3 SCOPE LOOP3.

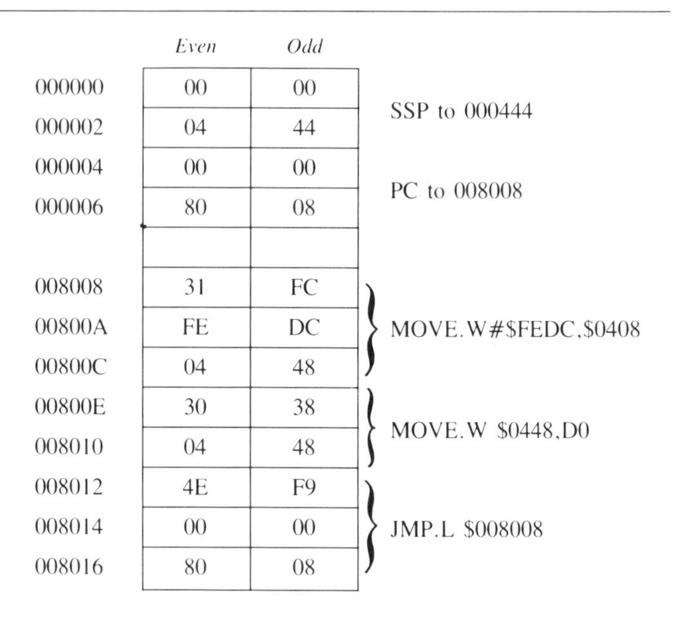

	Even	*Odd*	
000000	00	00	SSP to 000444
000002	04	44	
000004	00	00	PC to 008008
000006	80	08	
008008	31	FC	MOVE.W#$FEDC,$0408
00800A	FE	DC	
00800C	04	48	
00800E	30	38	MOVE.W $0448,D0
008010	04	48	
008012	4E	F9	JMP.L $008008
008014	00	00	
008016	80	08	

APPENDIX B TIMING DIAGRAMS

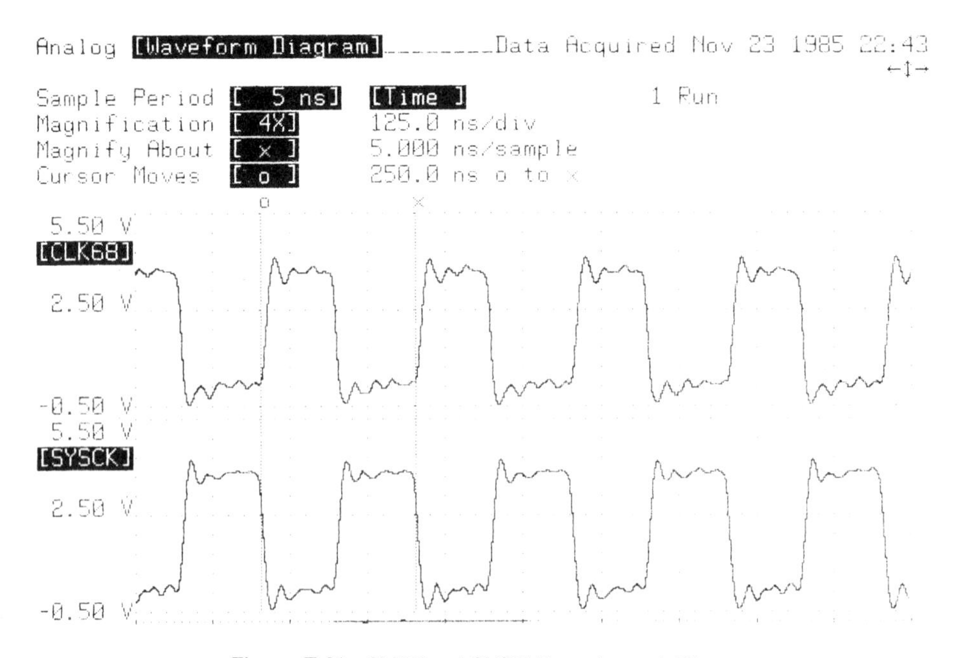

Figure T-01 CLK68 and SYSCLK running at 4 Mhz.

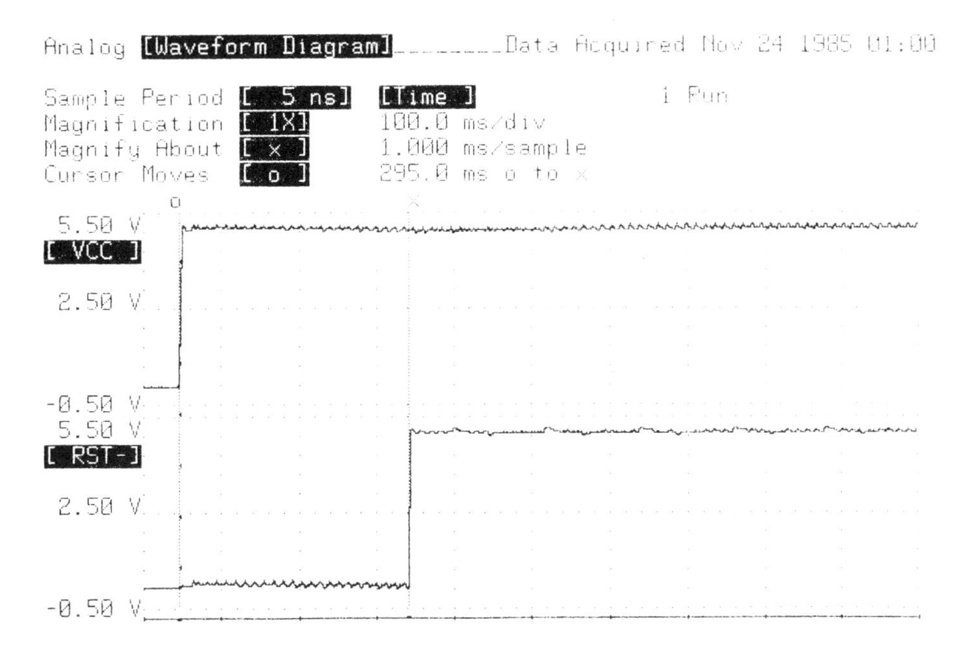

Figure T-02 Power-up reset sequence.

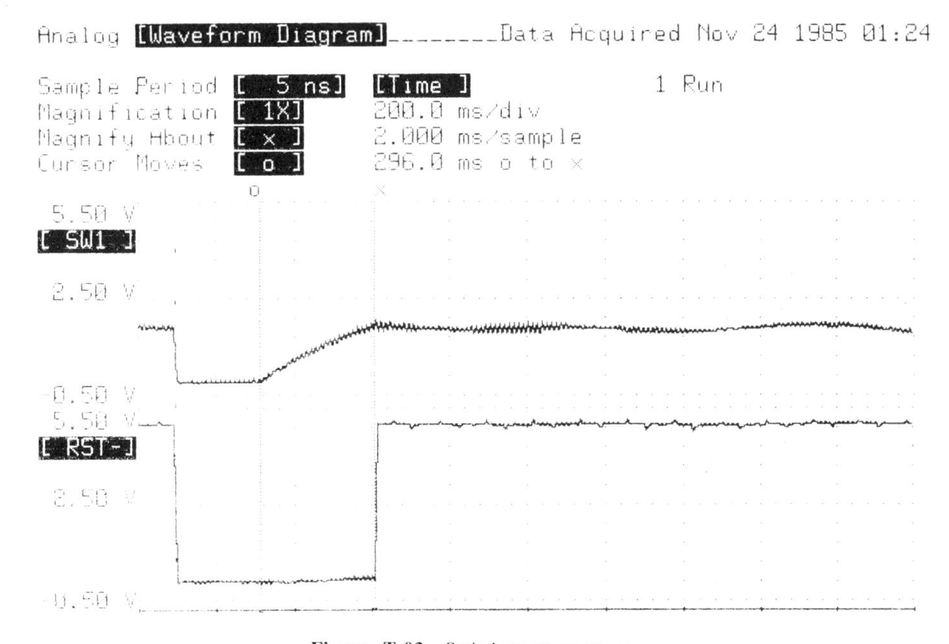

Figure T-03 Switch reset sequence.

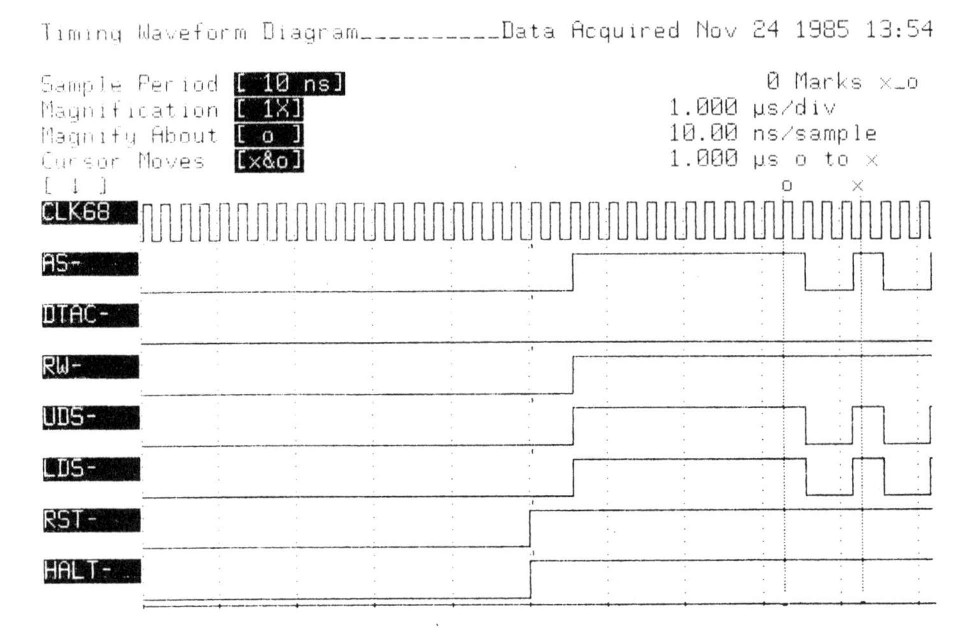

Figure T-04 Power-up sequence on freerun.

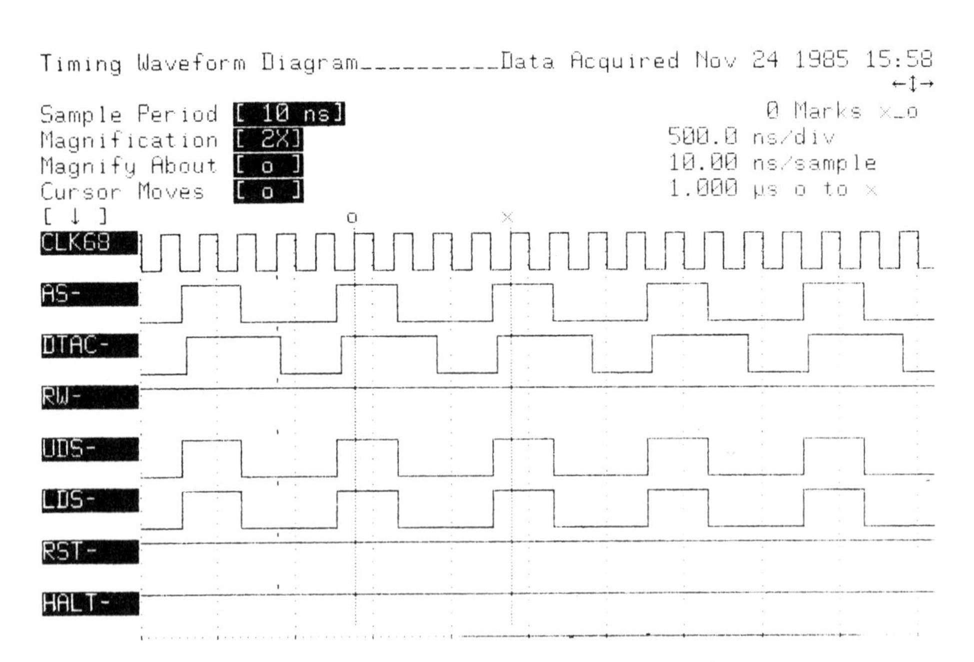

Figure T-05 Freerun with 0 waits.

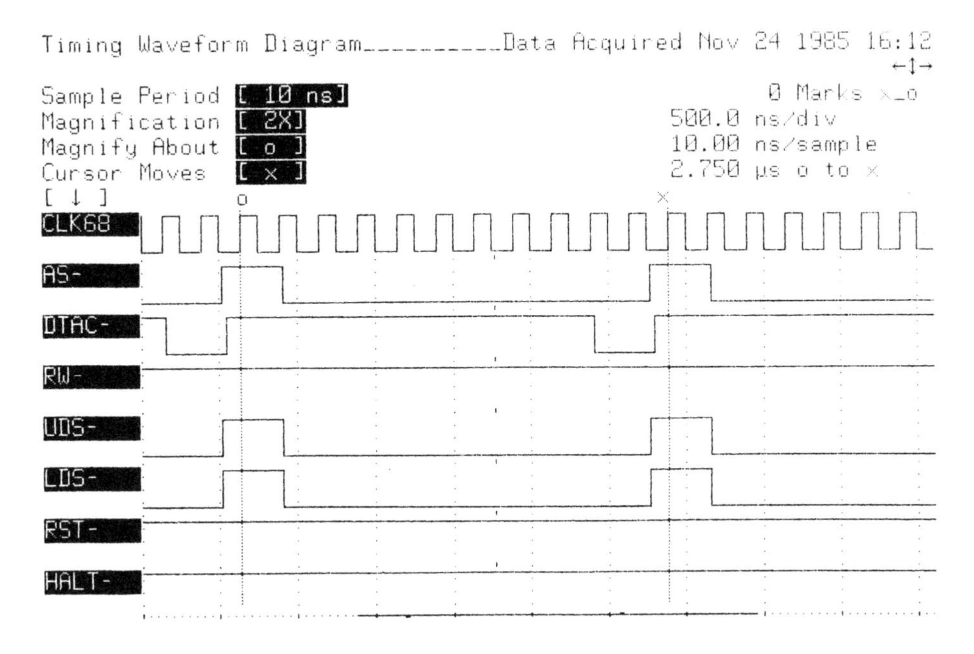

Figure T-06 Freerun with 7 waits.

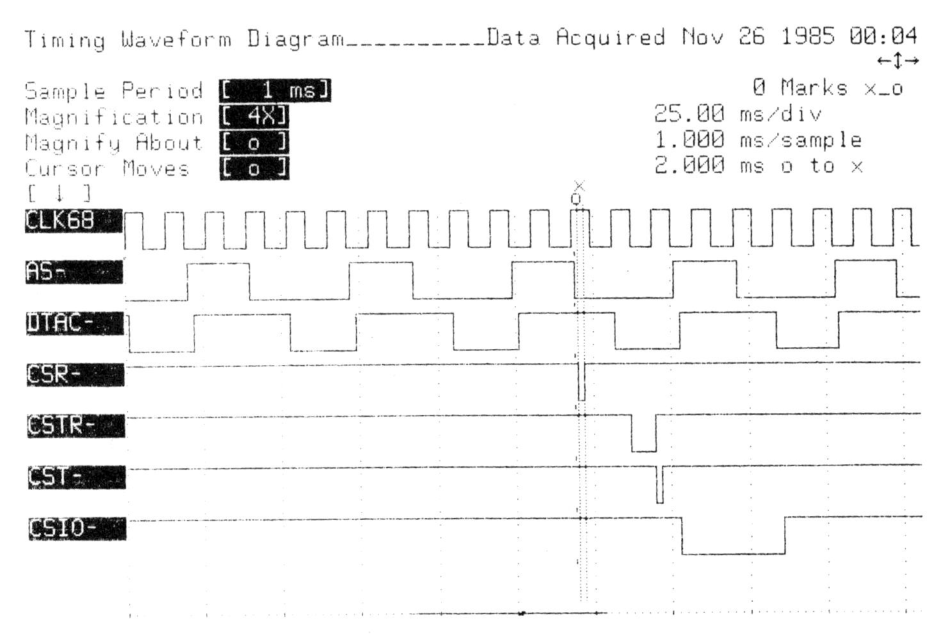

Figure T-07 Decoder module outputs on freerun.

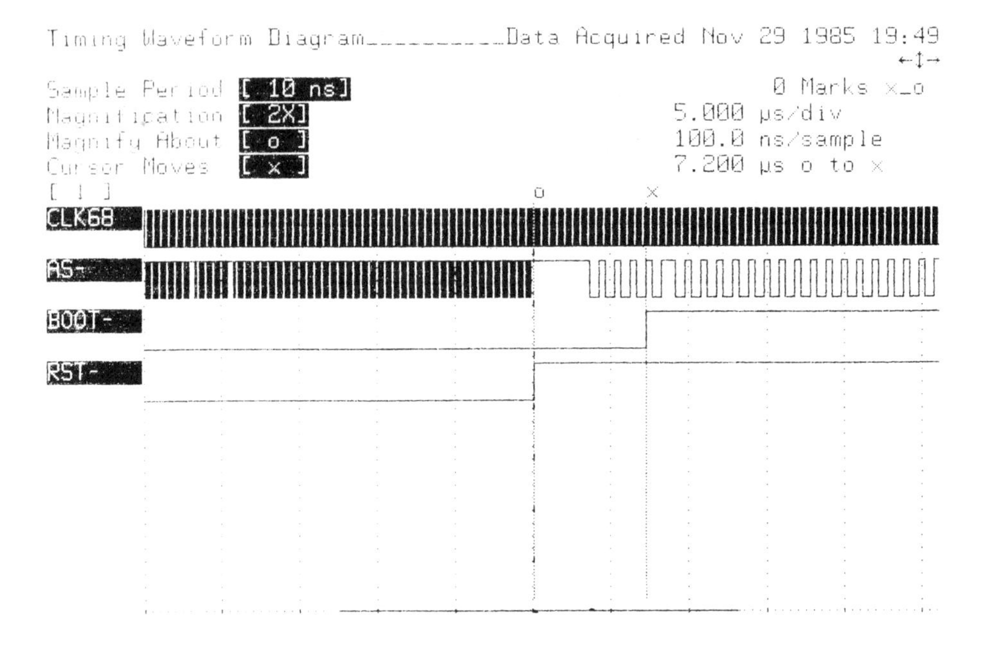

Figure T-08 BOOT* vs RESET* on power-up.

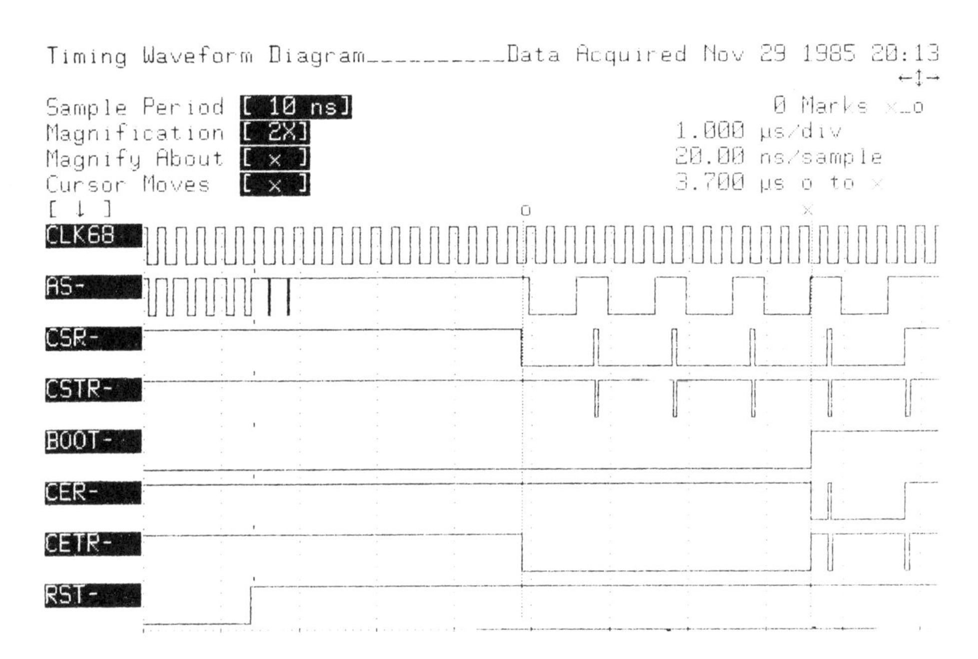

Figure T-09 Chip enables on power-up.

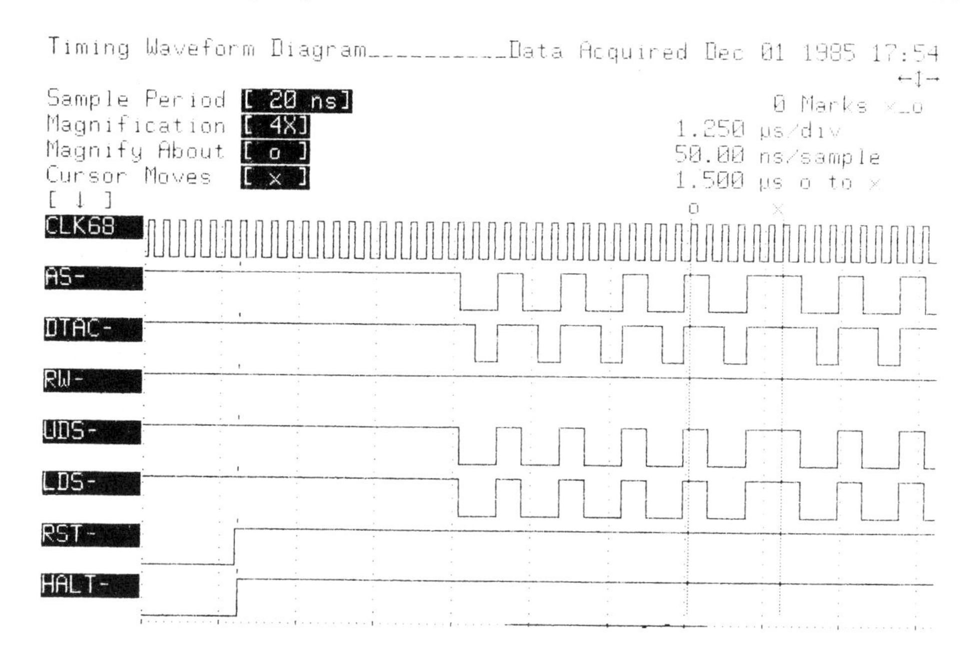

Figure T-10 TUTOR start-up sequence.

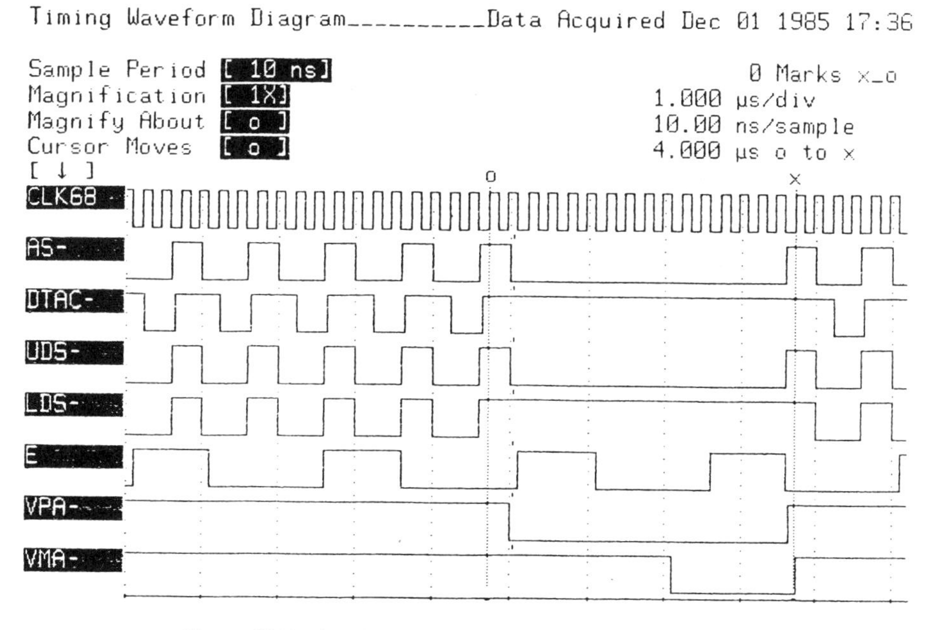

Figure T-11 Synchronous read bus cycle during TUTOR run.

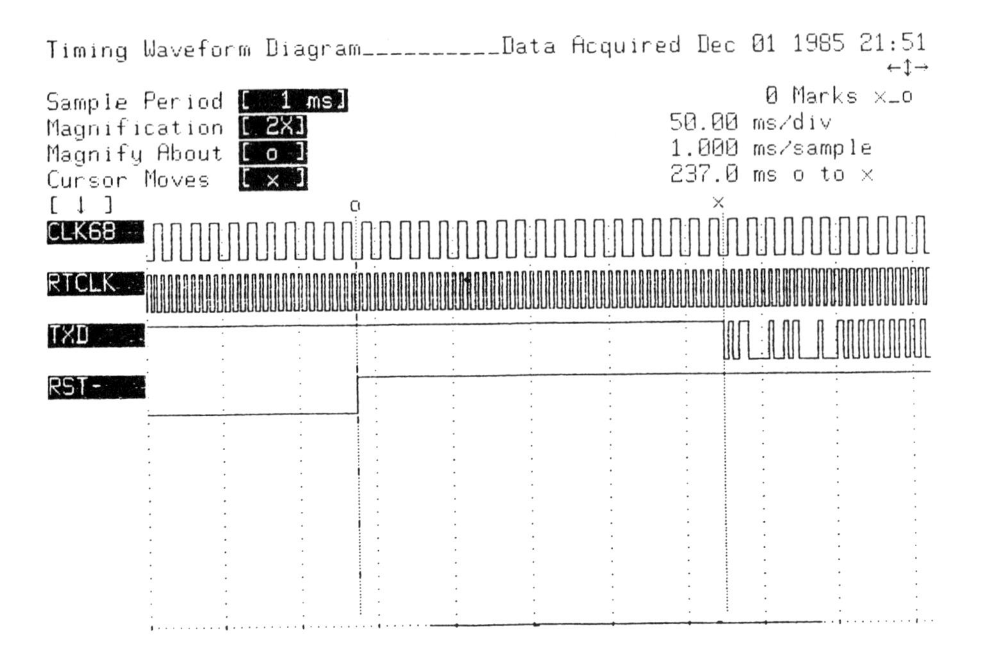

Figure T-12 $T_X D$ vs RESET* on power-up.

APPENDIX C SCHEMATICS

Drawing Index

Description	*Drawing #*
Power/Ground Table	S-00
Parts Location Diagram	S-01
Test Module	S-02
Clock Module	S-03
Reset Module	S-04
Freerun Module	S-05
DTACK* Module	S-06
Decoder Module	S-07
Test EPROM Module	S-08
Boot Module	S-09
Single-Step Module	S-10
RAM Module	S-11
TUTOR Module	S-12
I/O Module	S-13
Schematic Diagram	S-14

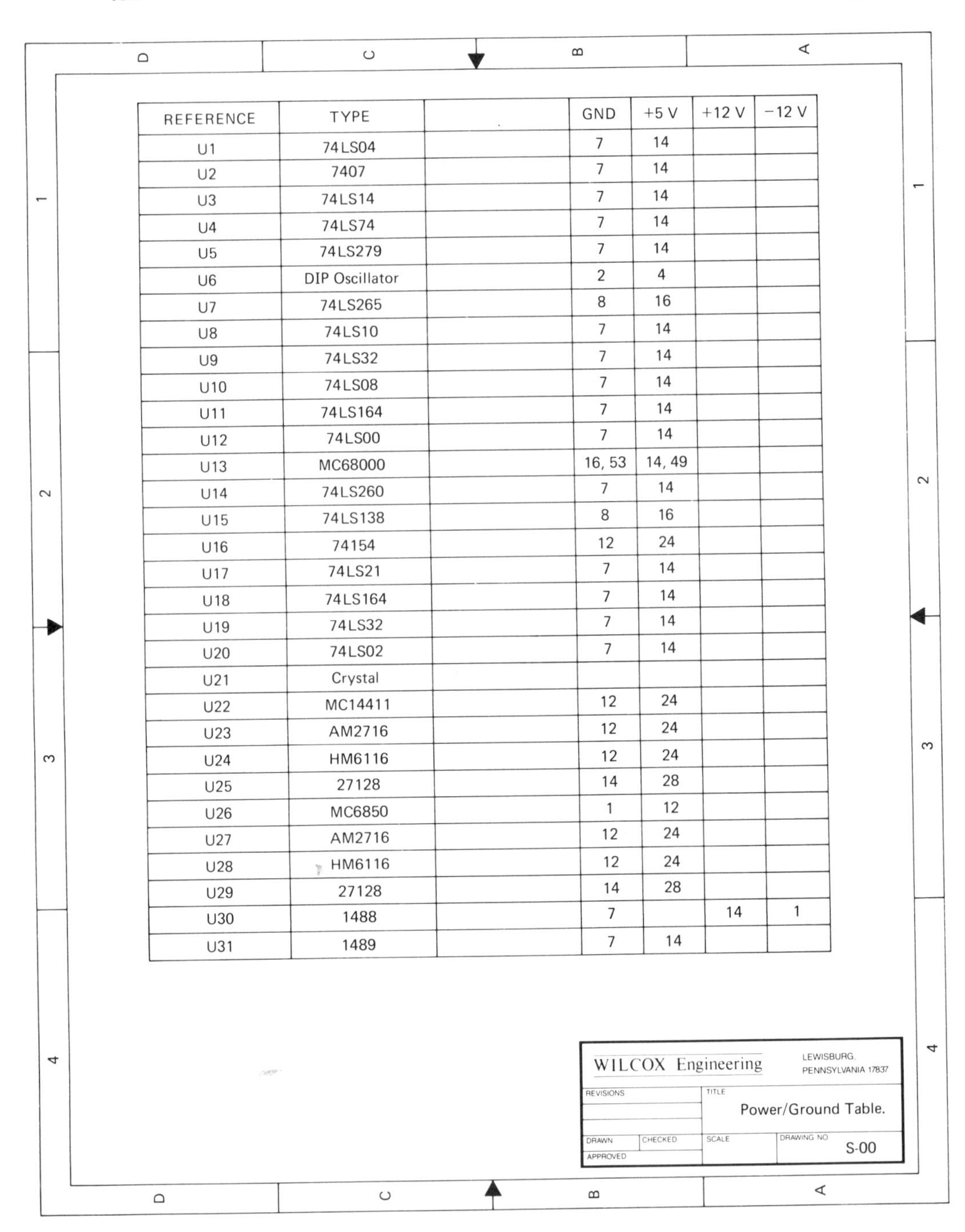

REFERENCE	TYPE		GND	+5 V	+12 V	−12 V
U1	74LS04		7	14		
U2	7407		7	14		
U3	74LS14		7	14		
U4	74LS74		7	14		
U5	74LS279		7	14		
U6	DIP Oscillator		2	4		
U7	74LS265		8	16		
U8	74LS10		7	14		
U9	74LS32		7	14		
U10	74LS08		7	14		
U11	74LS164		7	14		
U12	74LS00		7	14		
U13	MC68000		16, 53	14, 49		
U14	74LS260		7	14		
U15	74LS138		8	16		
U16	74154		12	24		
U17	74LS21		7	14		
U18	74LS164		7	14		
U19	74LS32		7	14		
U20	74LS02		7	14		
U21	Crystal					
U22	MC14411		12	24		
U23	AM2716		12	24		
U24	HM6116		12	24		
U25	27128		14	28		
U26	MC6850		1	12		
U27	AM2716		12	24		
U28	HM6116		12	24		
U29	27128		14	28		
U30	1488		7		14	1
U31	1489		7	14		

WILCOX Engineering — LEWISBURG, PENNSYLVANIA 17837

REVISIONS | TITLE: Power/Ground Table.

DRAWN | CHECKED | SCALE | DRAWING NO: S-00

APPROVED

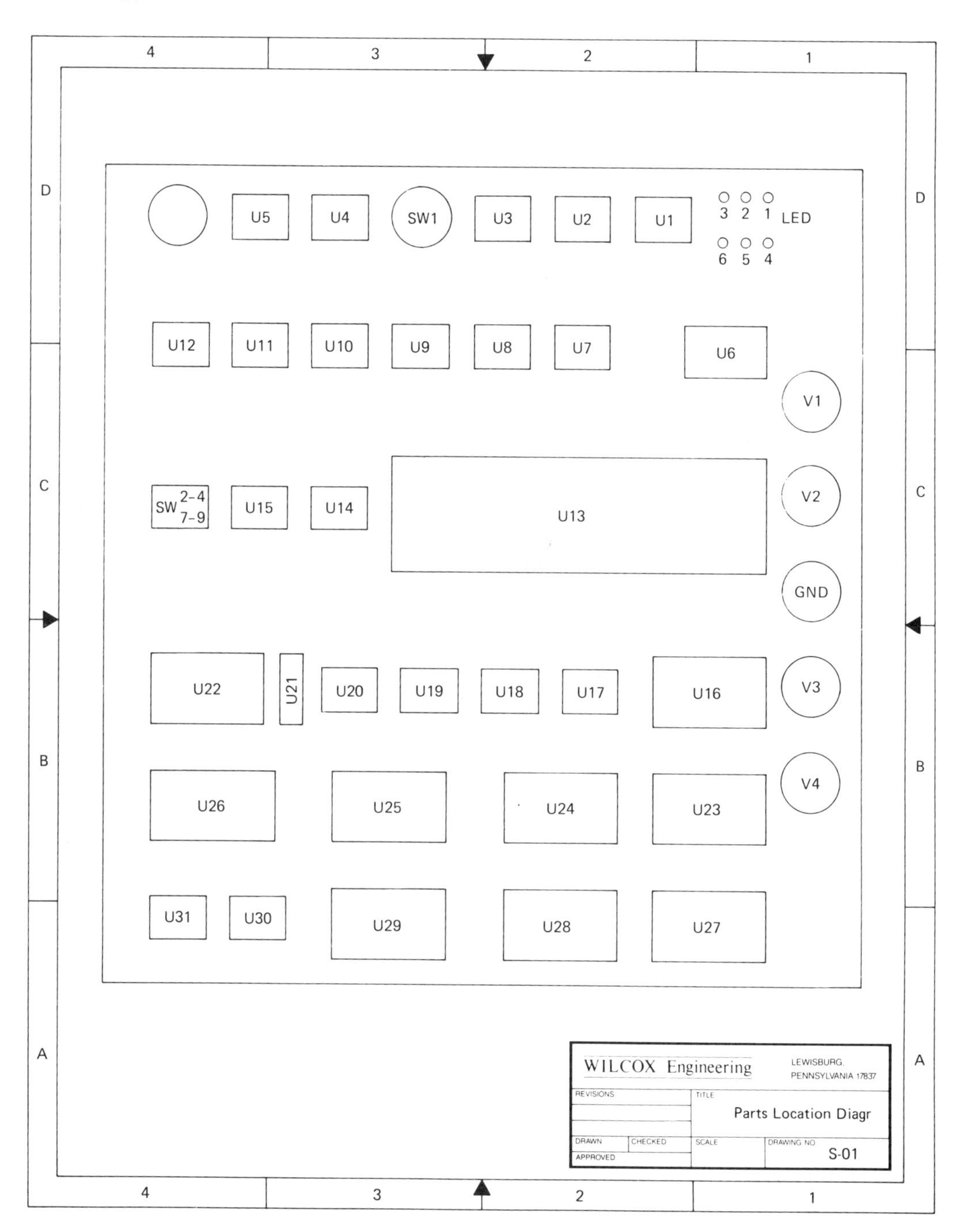
4
3
2
1
D
C
B
A
U5
U4
SW1
U3
U2
U1
3 2 1 LED
6 5 4
U12
U11
U10
U9
U8
U7
U6
V1
SW 2-4 7-9
U15
U14
U13
V2
GND
U22
U21
U20
U19
U18
U17
U16
V3
U26
U25
U24
U23
V4
U31
U30
U29
U28
U27
WILCOX Engineering
LEWISBURG, PENNSYLVANIA 17837
REVISIONS
TITLE
Parts Location Diagr
DRAWN
CHECKED
SCALE
DRAWING NO
APPROVED
S-01

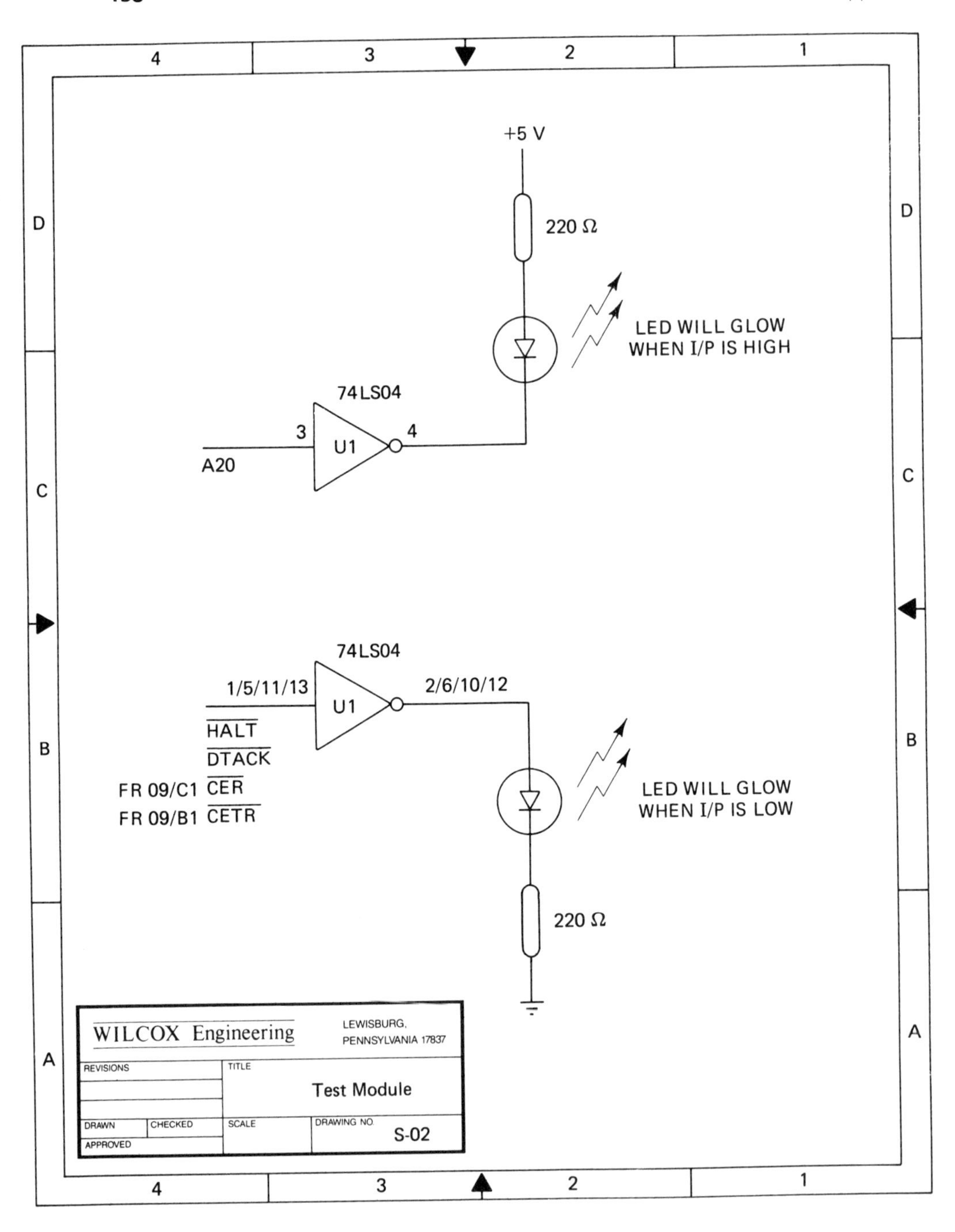
4
3
2
1
D
C
B
A
+5 V
220 Ω
LED WILL GLOW
WHEN I/P IS HIGH
74LS04
3
4
U1
A20
74LS04
1/5/11/13
2/6/10/12
U1
HALT
DTACK
FR 09/C1 CER
FR 09/B1 CETR
LED WILL GLOW
WHEN I/P IS LOW
220 Ω
WILCOX Engineering
LEWISBURG, PENNSYLVANIA 17837
REVISIONS
TITLE
Test Module
DRAWN
CHECKED
SCALE
DRAWING NO.
S-02
APPROVED

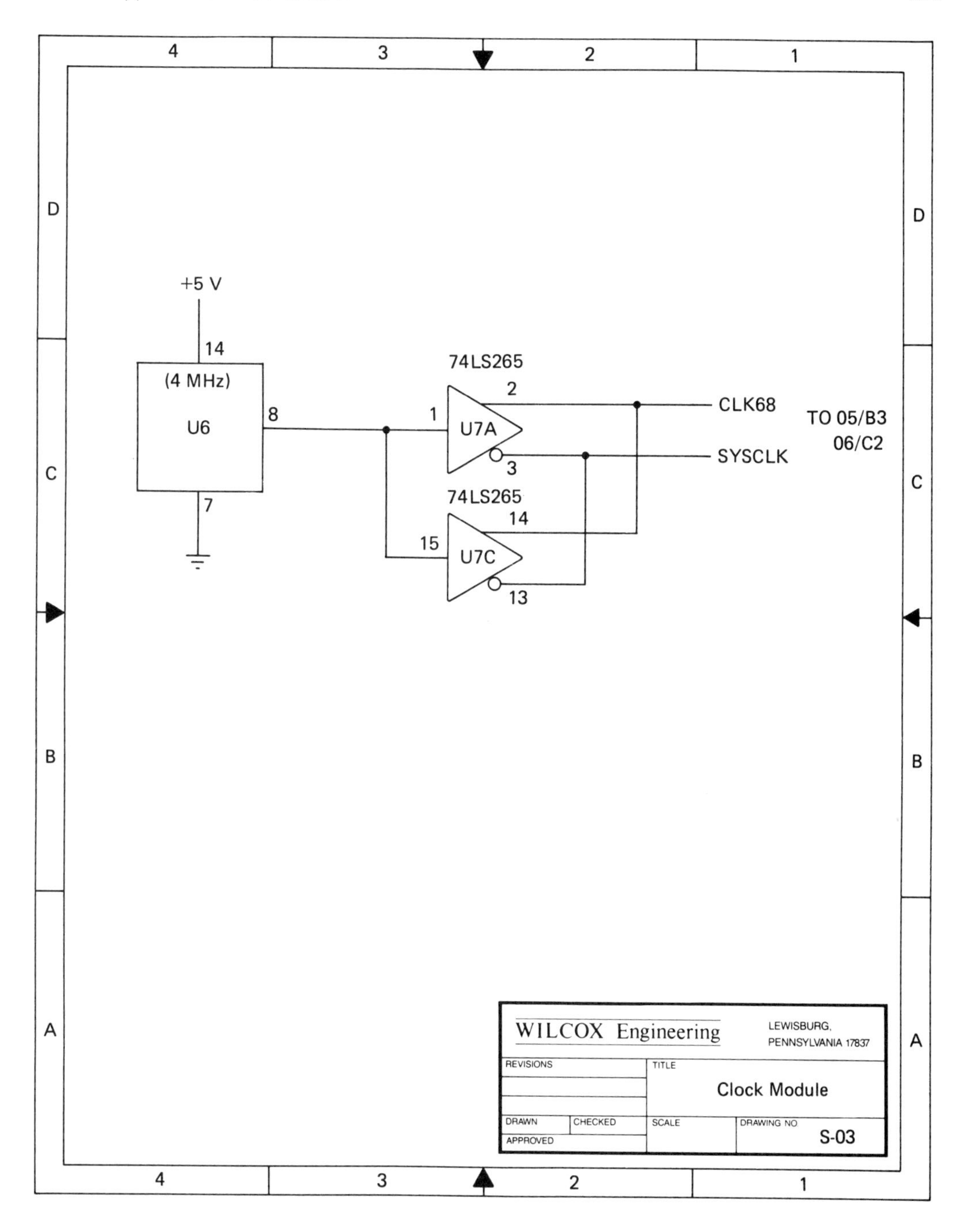

4
3
2
1
D
C
B
A
+5 V
14
(4 MHz)
U6
8
7
74LS265
2
1
U7A
3
CLK68
SYSCLK
TO 05/B3
06/C2
74LS265
14
15
U7C
13
WILCOX Engineering
LEWISBURG, PENNSYLVANIA 17837
REVISIONS
TITLE
Clock Module
DRAWN
CHECKED
SCALE
DRAWING NO.
S-03
APPROVED

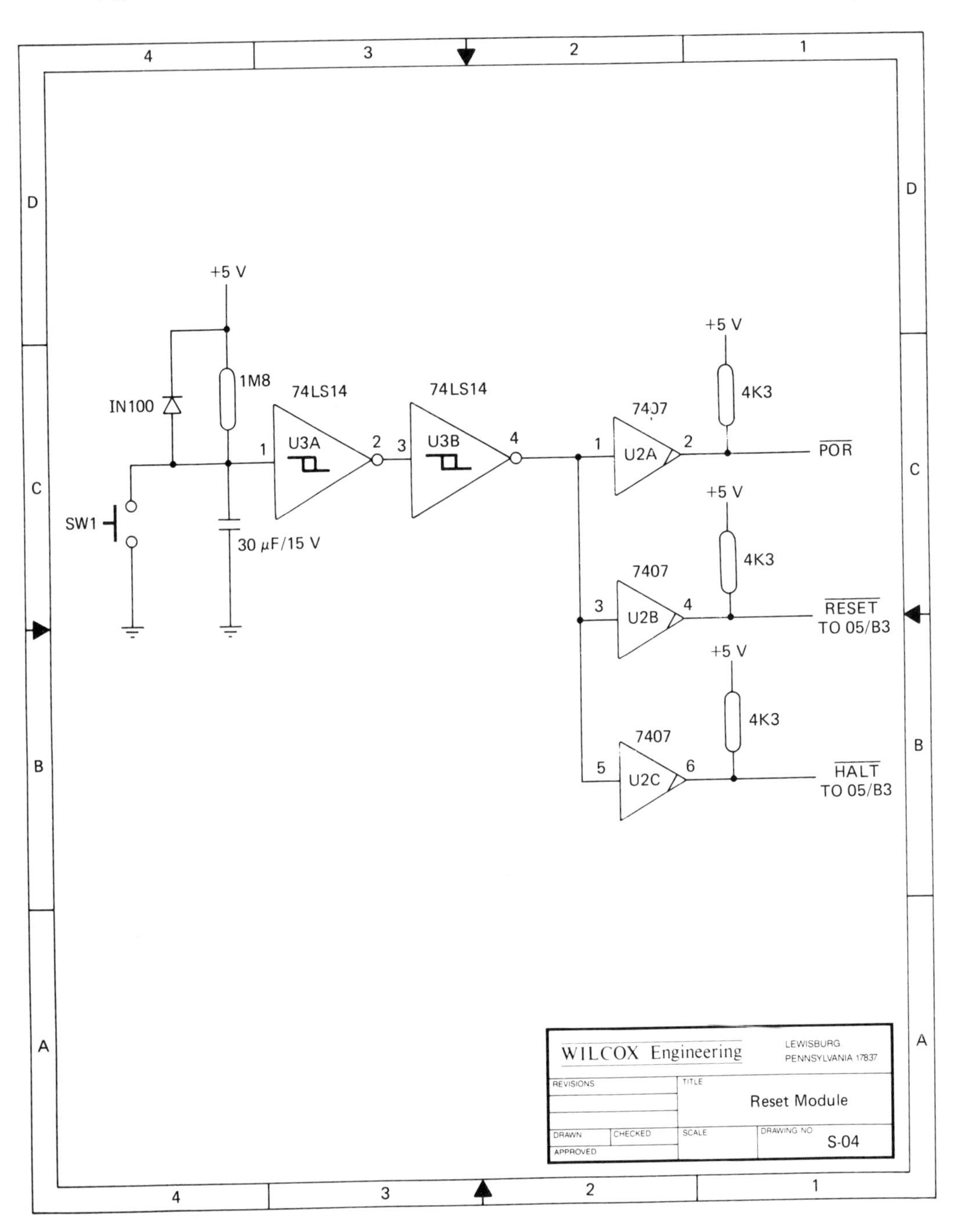
+5 V
1M8
IN100
74LS14
U3A
74LS14
U3B
SW1
30 μF/15 V
+5 V
4K3
7407
U2A
POR
+5 V
4K3
7407
U2B
RESET
TO 05/B3
+5 V
4K3
7407
U2C
HALT
TO 05/B3
WILCOX Engineering
LEWISBURG, PENNSYLVANIA 17837
REVISIONS
TITLE
Reset Module
DRAWN
CHECKED
SCALE
DRAWING NO
S-04
APPROVED

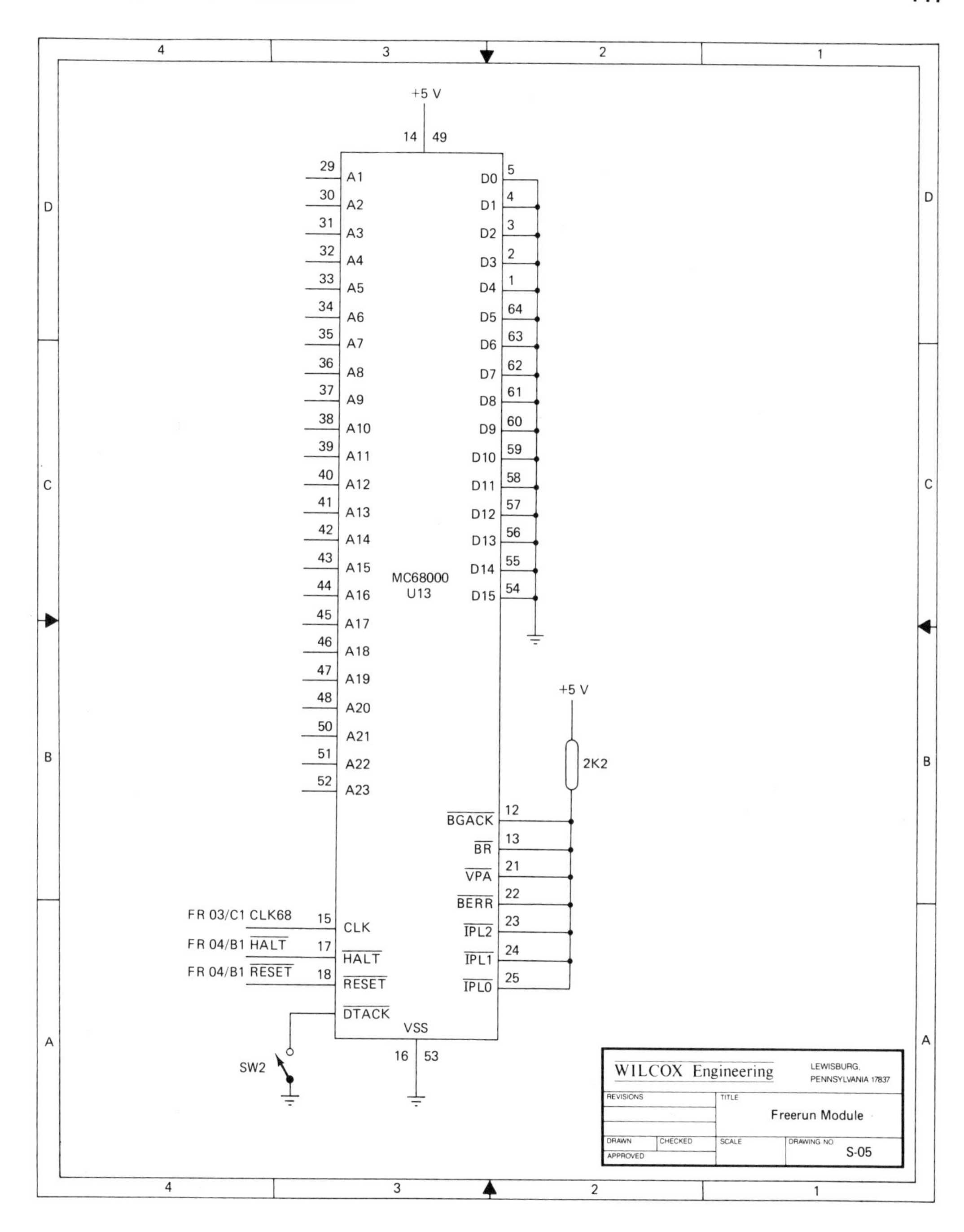

+5 V
14 49
MC68000
U13
A1 A2 A3 A4 A5 A6 A7 A8 A9 A10 A11 A12 A13 A14 A15 A16 A17 A18 A19 A20 A21 A22 A23
D0 D1 D2 D3 D4 D5 D6 D7 D8 D9 D10 D11 D12 D13 D14 D15
+5 V
2K2
BGACK
BR
VPA
BERR
IPL2
IPL1
IPL0
FR 03/C1 CLK68
FR 04/B1 HALT
FR 04/B1 RESET
CLK
HALT
RESET
DTACK
VSS
16 53
SW2
WILCOX Engineering
LEWISBURG, PENNSYLVANIA 17837
REVISIONS
TITLE
Freerun Module
DRAWN
CHECKED
SCALE
DRAWING NO
S-05
APPROVED

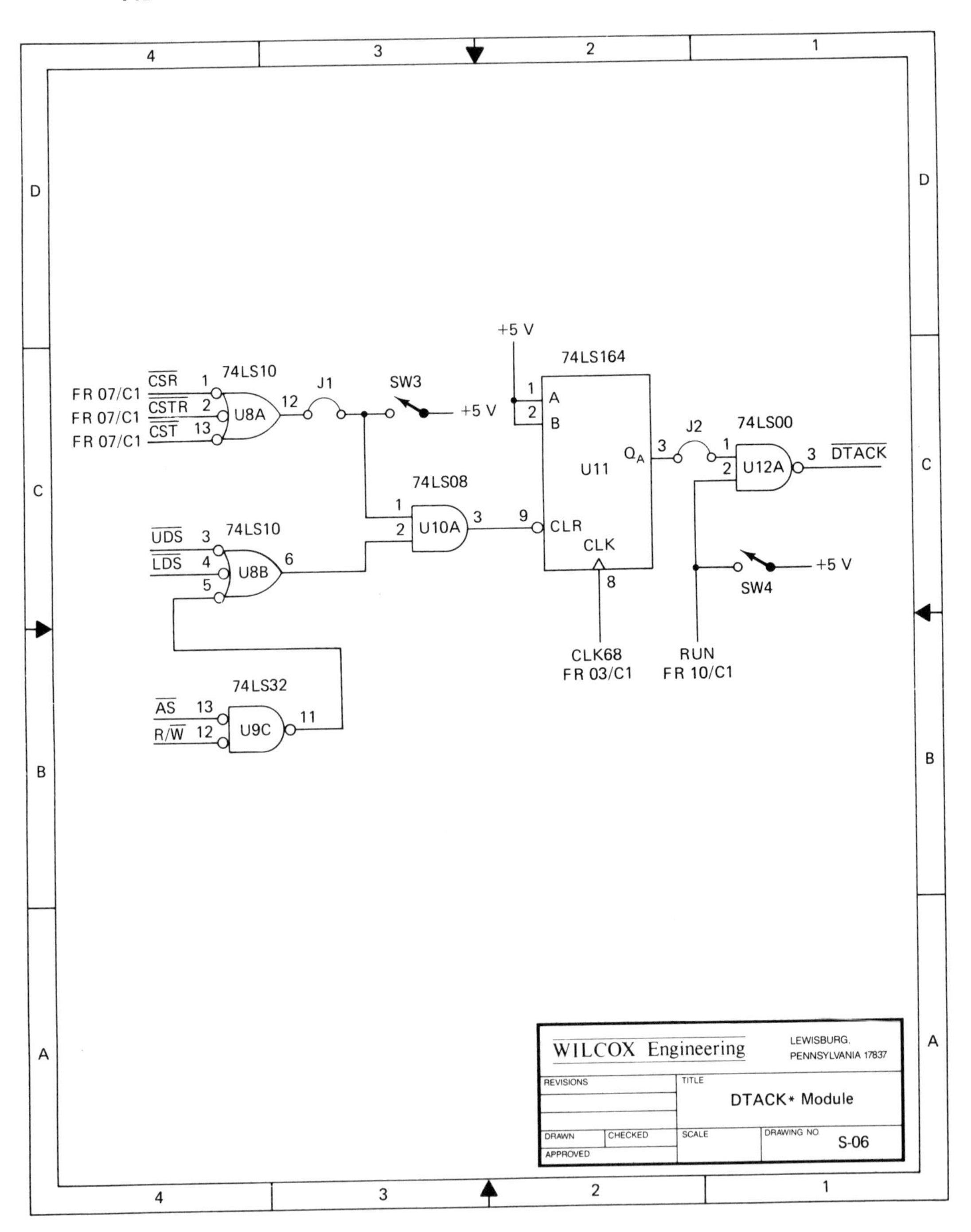

4
3
2
1
D
C
B
A
CSR
CSTR
CST
FR 07/C1
FR 07/C1
FR 07/C1
74LS10
U8A
J1
SW3
+5 V
74LS164
A
B
U11
Q_A
CLR
CLK
J2
74LS00
U12A
DTACK
74LS08
U10A
UDS
LDS
74LS10
U8B
SW4
CLK68
FR 03/C1
RUN
FR 10/C1
74LS32
AS
R/W
U9C
WILCOX Engineering
LEWISBURG, PENNSYLVANIA 17837
REVISIONS
TITLE
DTACK* Module
DRAWN
CHECKED
SCALE
DRAWING NO
S-06
APPROVED

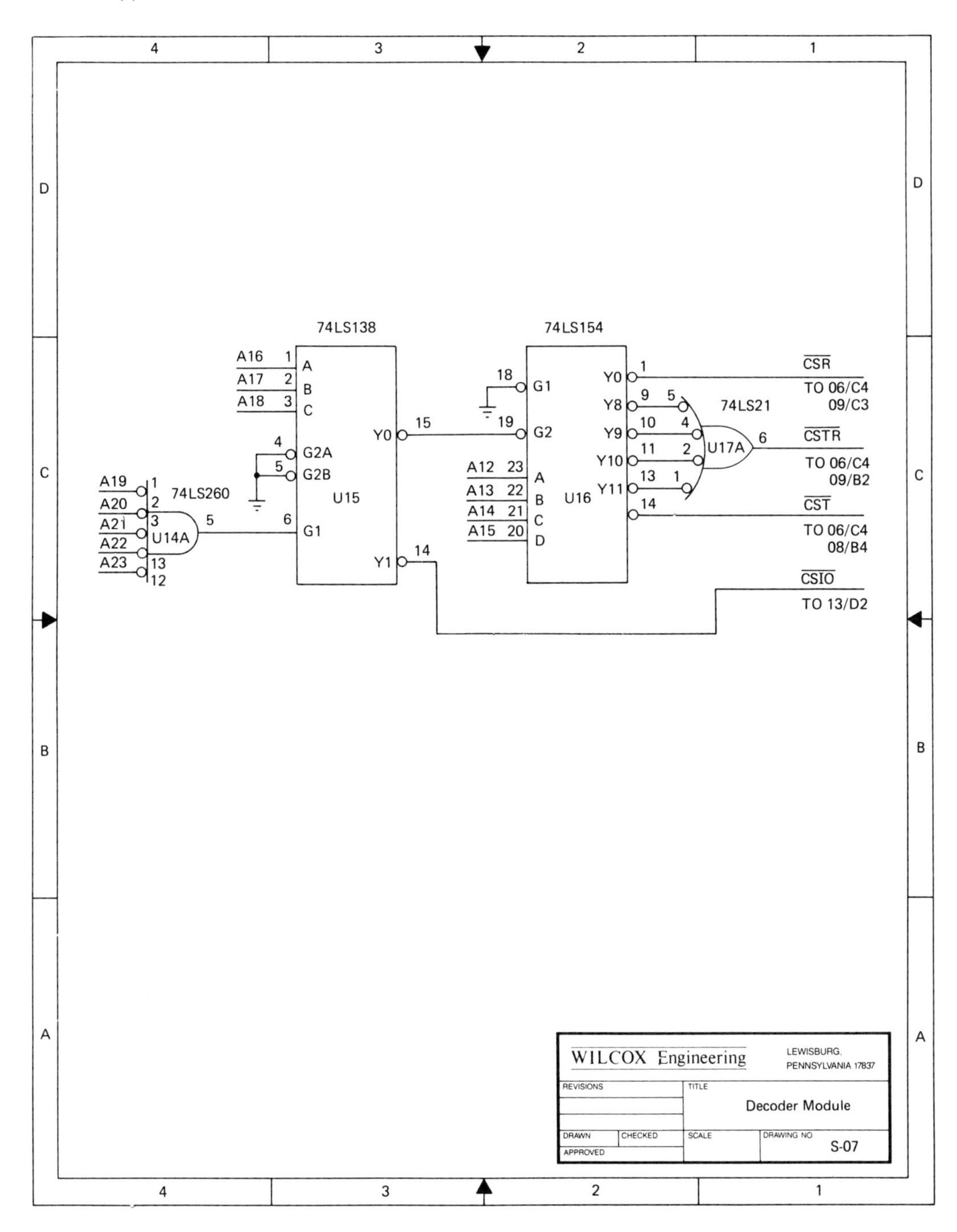
74LS138
74LS154
A16
A17
A18
A
B
C
G2A
G2B
G1
U15
Y0
Y1
G1
G2
A12
A13
A14
A15
A
B
C
D
U16
Y0
Y8
Y9
Y10
Y11
74LS21
U17A
A19
A20
A21
A22
A23
74LS260
U14A
CSR
TO 06/C4
09/C3
CSTR
TO 06/C4
09/B2
CST
TO 06/C4
08/B4
CSIO
TO 13/D2
WILCOX Engineering
LEWISBURG, PENNSYLVANIA 17837
REVISIONS
TITLE
Decoder Module
DRAWN
CHECKED
SCALE
DRAWING NO
S-07
APPROVED

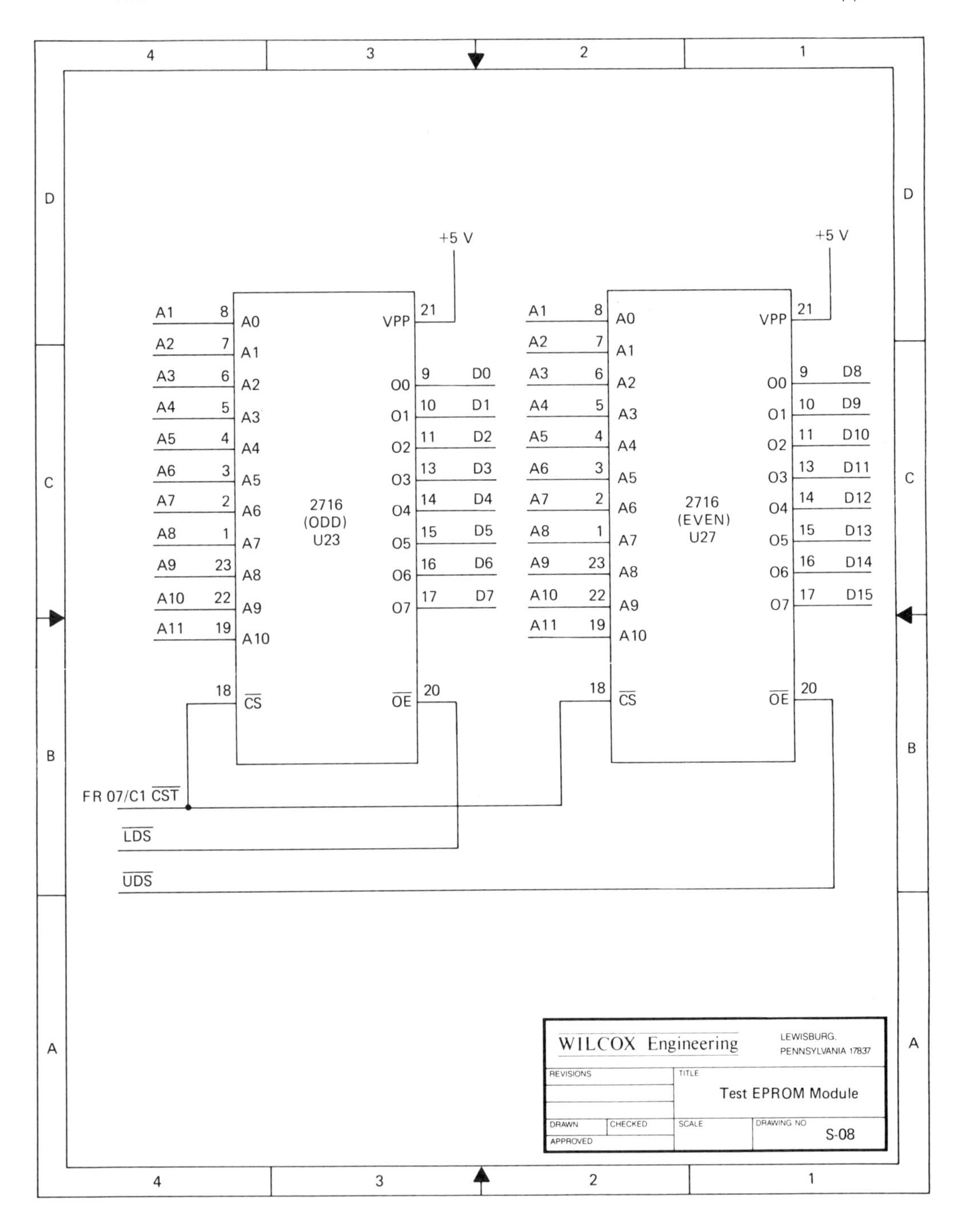

4
3
2
1
D
C
B
A
+5 V
A1 8 A0
A2 7 A1
A3 6 A2
A4 5 A3
A5 4 A4
A6 3 A5
A7 2 A6
A8 1 A7
A9 23 A8
A10 22 A9
A11 19 A10
VPP 21
O0 9 D0
O1 10 D1
O2 11 D2
O3 13 D3
O4 14 D4
O5 15 D5
O6 16 D6
O7 17 D7
2716
(ODD)
U23
18 CS
OE 20
2716
(EVEN)
U27
O0 9 D8
O1 10 D9
O2 11 D10
O3 13 D11
O4 14 D12
O5 15 D13
O6 16 D14
O7 17 D15
FR 07/C1 CST
LDS
UDS
WILCOX Engineering
LEWISBURG, PENNSYLVANIA 17837
REVISIONS
TITLE
Test EPROM Module
DRAWN
CHECKED
SCALE
DRAWING NO
S-08
APPROVED

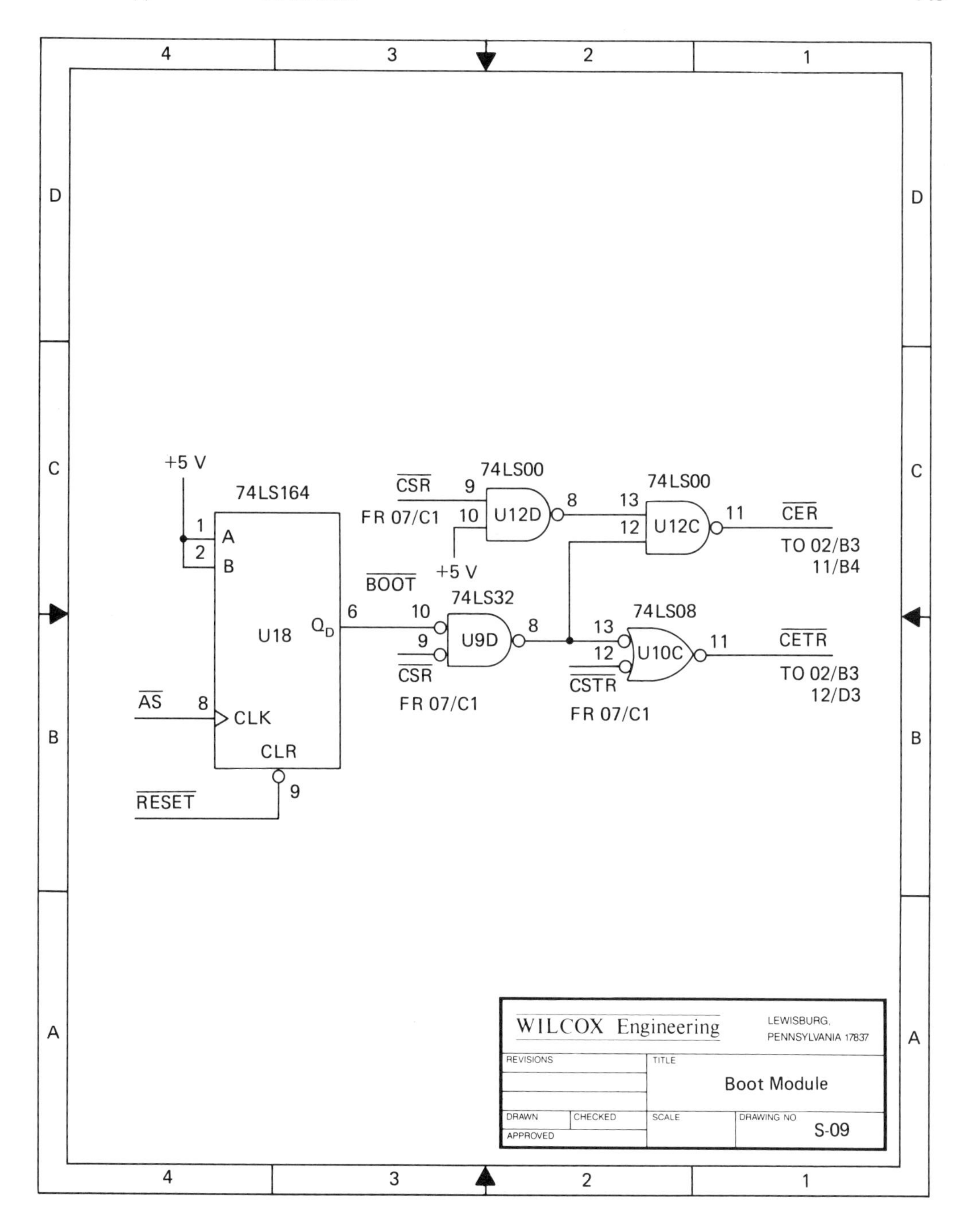

4
3
2
1
D
C
B
A
+5 V
74LS164
1
A
2
B
U18
Q_D
6
$\overline{AS}$
8
CLK
CLR
9
$\overline{RESET}$
$\overline{CSR}$
9
74LS00
FR 07/C1
10
U12D
8
+5 V
$\overline{BOOT}$
74LS32
10
9
U9D
8
$\overline{CSR}$
FR 07/C1
74LS00
13
12
U12C
11
$\overline{CER}$
TO 02/B3
11/B4
74LS08
13
12
U10C
11
$\overline{CSTR}$
FR 07/C1
$\overline{CETR}$
TO 02/B3
12/D3
WILCOX Engineering
LEWISBURG, PENNSYLVANIA 17837
REVISIONS
TITLE
Boot Module
DRAWN
CHECKED
SCALE
DRAWING NO
S-09
APPROVED

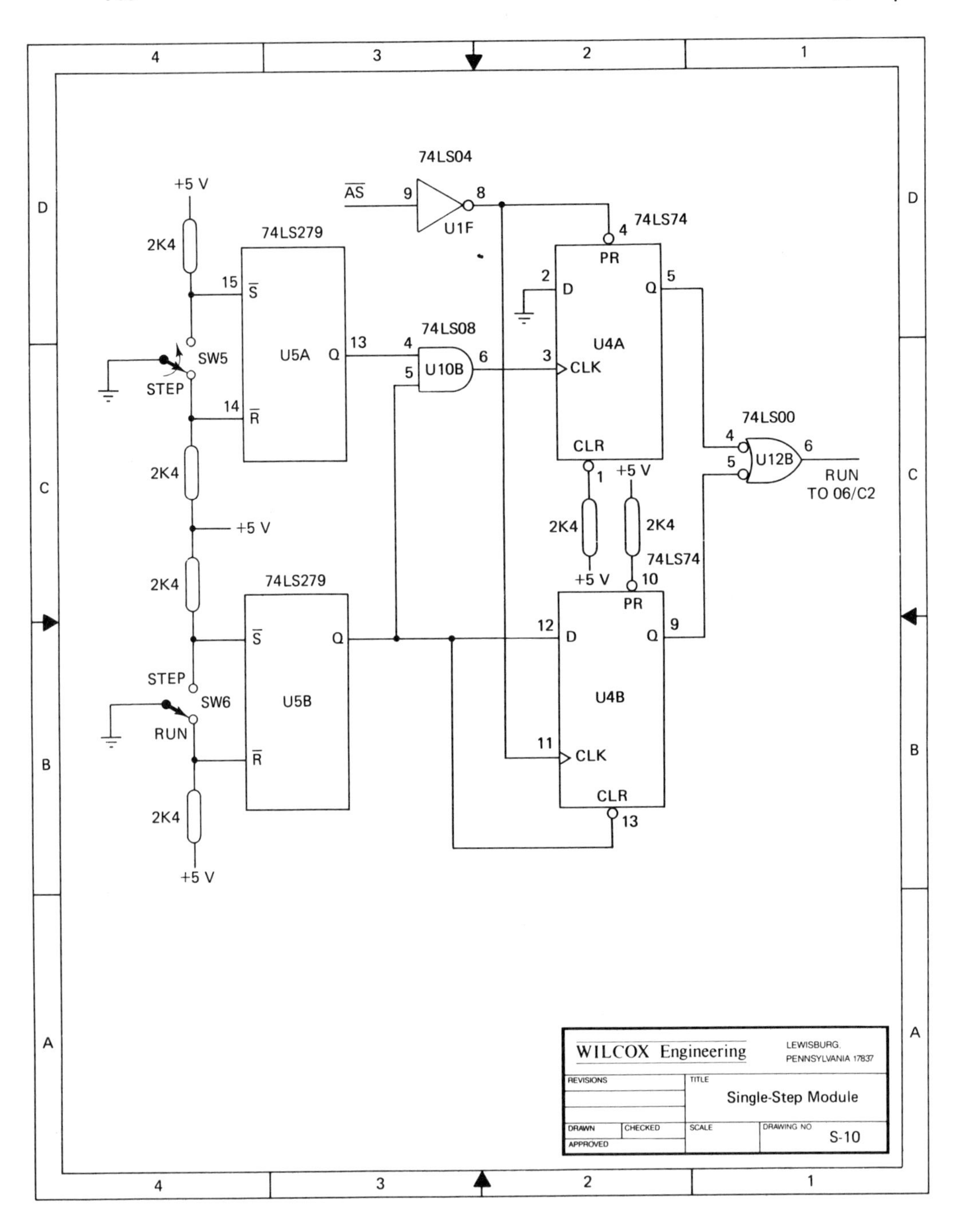
74LS04
AS
9
8
U1F
74LS74
4
PR
+5 V
2K4
74LS279
15
S
SW5
STEP
14
R
U5A
Q
13
74LS08
4
5
U10B
6
2
D
Q
5
3
CLK
U4A
CLR
1
+5 V
2K4
2K4
74LS00
4
5
U12B
6
RUN
TO 06/C2
+5 V
2K4
74LS279
S
Q
STEP
SW6
RUN
U5B
R
2K4
+5 V
74LS74
10
PR
12
D
Q
9
U4B
11
CLK
CLR
13
WILCOX Engineering
LEWISBURG,
PENNSYLVANIA 17837
REVISIONS
TITLE
Single-Step Module
DRAWN
CHECKED
SCALE
DRAWING NO
S-10
APPROVED

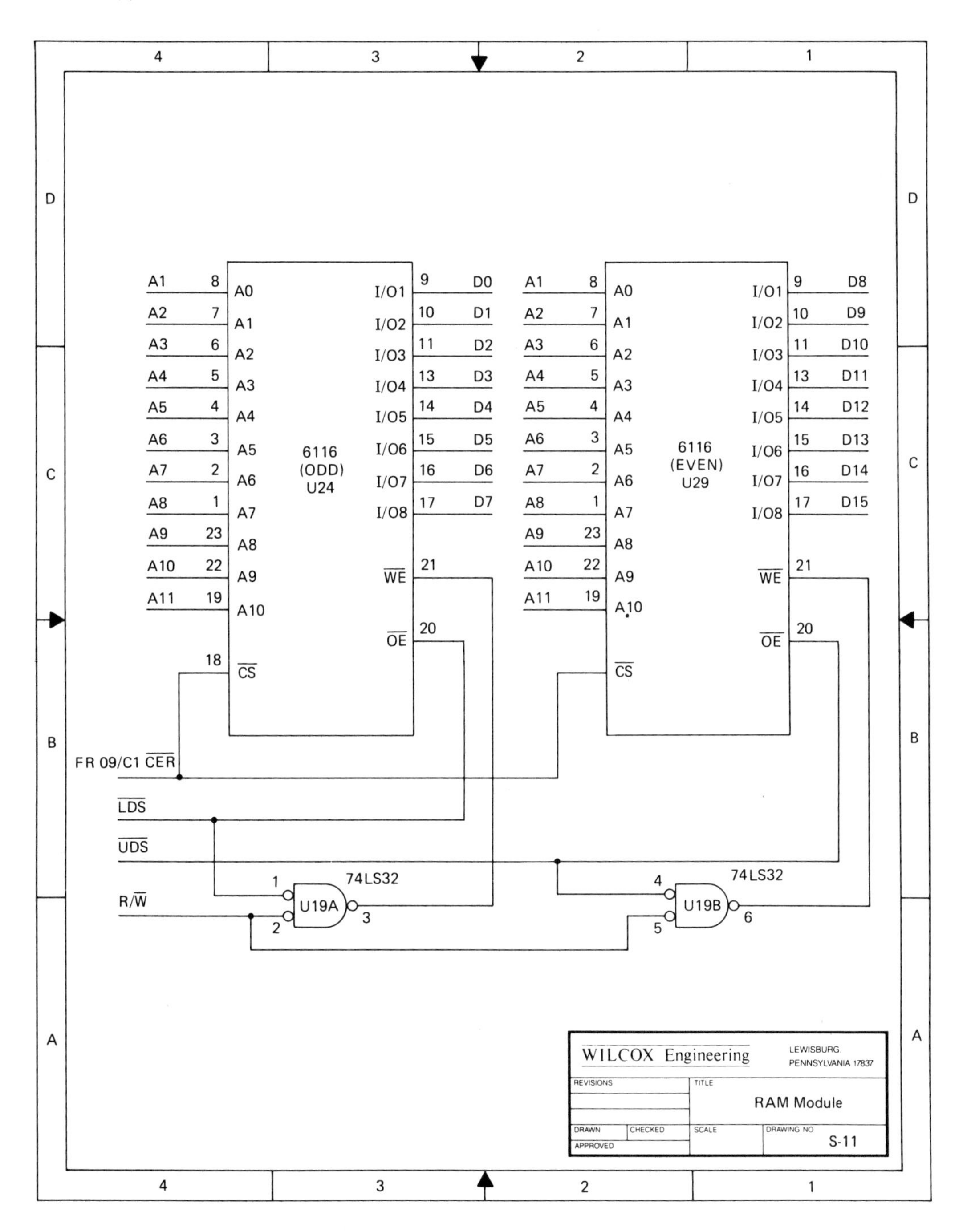

4
3
2
1
D
C
B
A
A1 8 A0
A2 7 A1
A3 6 A2
A4 5 A3
A5 4 A4
A6 3 A5
A7 2 A6
A8 1 A7
A9 23 A8
A10 22 A9
A11 19 A10
18 CS
6116 (ODD) U24
I/O1 9 D0
I/O2 10 D1
I/O3 11 D2
I/O4 13 D3
I/O5 14 D4
I/O6 15 D5
I/O7 16 D6
I/O8 17 D7
WE 21
OE 20
6116 (EVEN) U29
I/O1 9 D8
I/O2 10 D9
I/O3 11 D10
I/O4 13 D11
I/O5 14 D12
I/O6 15 D13
I/O7 16 D14
I/O8 17 D15
FR 09/C1 CER
LDS
UDS
R/W
74LS32
U19A
74LS32
U19B
WILCOX Engineering
LEWISBURG, PENNSYLVANIA 17837
REVISIONS
TITLE
RAM Module
DRAWN
CHECKED
SCALE
DRAWING NO
S-11
APPROVED

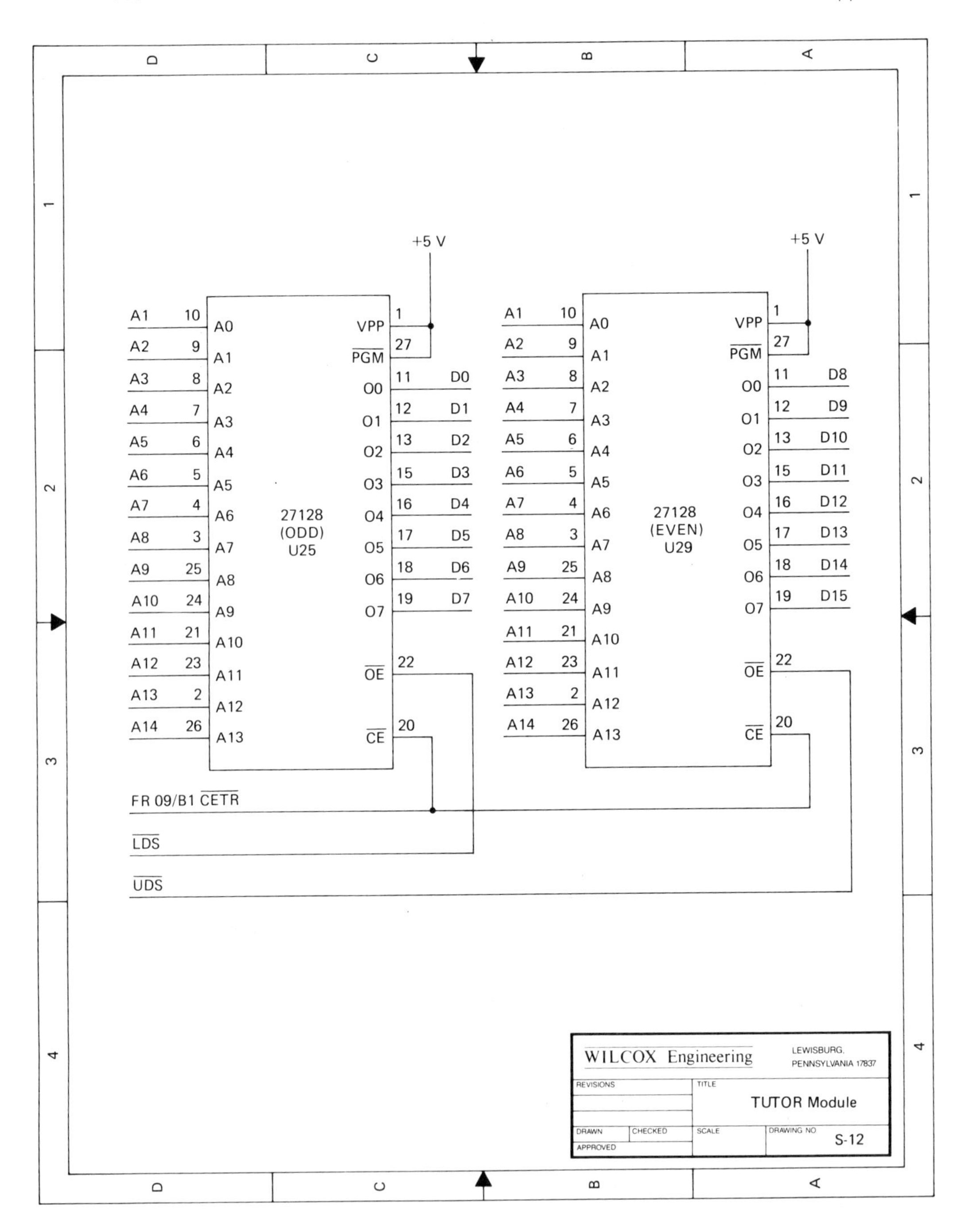
+5 V
A1 10 A0
A2 9 A1
A3 8 A2
A4 7 A3
A5 6 A4
A6 5 A5
A7 4 A6
A8 3 A7
A9 25 A8
A10 24 A9
A11 21 A10
A12 23 A11
A13 2 A12
A14 26 A13
VPP 1
PGM 27
O0 11 D0
O1 12 D1
O2 13 D2
O3 15 D3
O4 16 D4
O5 17 D5
O6 18 D6
O7 19 D7
OE 22
CE 20
27128
(ODD)
U25
27128
(EVEN)
U29
O0 11 D8
O1 12 D9
O2 13 D10
O3 15 D11
O4 16 D12
O5 17 D13
O6 18 D14
O7 19 D15
FR 09/B1 CETR
LDS
UDS
WILCOX Engineering
LEWISBURG, PENNSYLVANIA 17837
REVISIONS
TITLE
TUTOR Module
DRAWN
CHECKED
SCALE
DRAWING NO
S-12
APPROVED

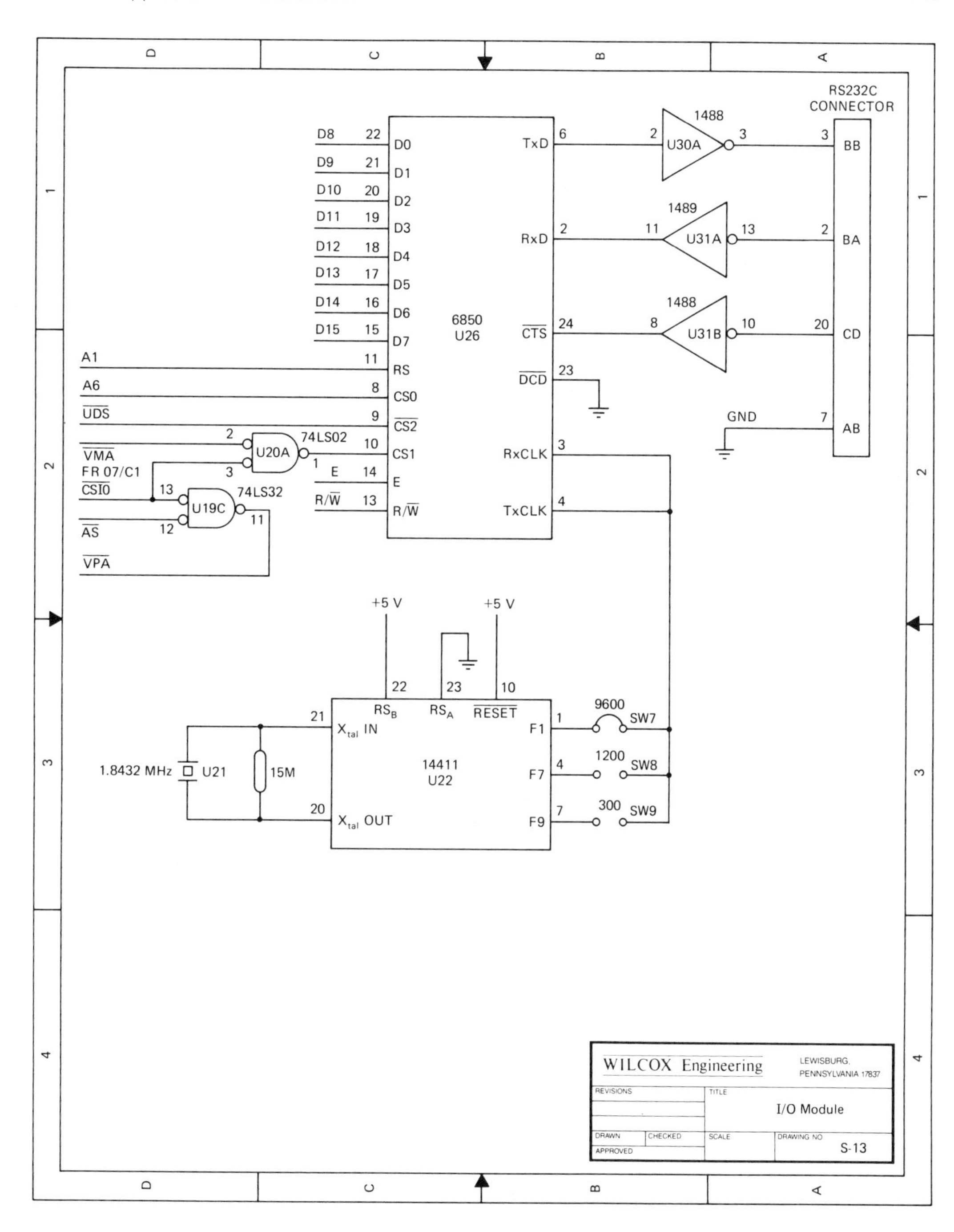
RS232C CONNECTOR
1488
U30A
1489
U31A
1488
U31B
TxD
RxD
CTS
DCD
RxCLK
TxCLK
6850
U26
D0 D1 D2 D3 D4 D5 D6 D7
D8 D9 D10 D11 D12 D13 D14 D15
A1
A6
UDS
VMA
FR 07/C1
CSIO
AS
VPA
RS
CS0
CS2
CS1
E
R/W
74LS02
U20A
74LS32
U19C
BB
BA
CD
AB
GND
+5 V
+5 V
RS_B
RS_A
RESET
X_tal IN
X_tal OUT
14411
U22
F1
F7
F9
9600
SW7
1200
SW8
300
SW9
1.8432 MHz
U21
15M
WILCOX Engineering
LEWISBURG, PENNSYLVANIA 17837
REVISIONS
TITLE
I/O Module
DRAWN
CHECKED
SCALE
DRAWING NO
S-13
APPROVED

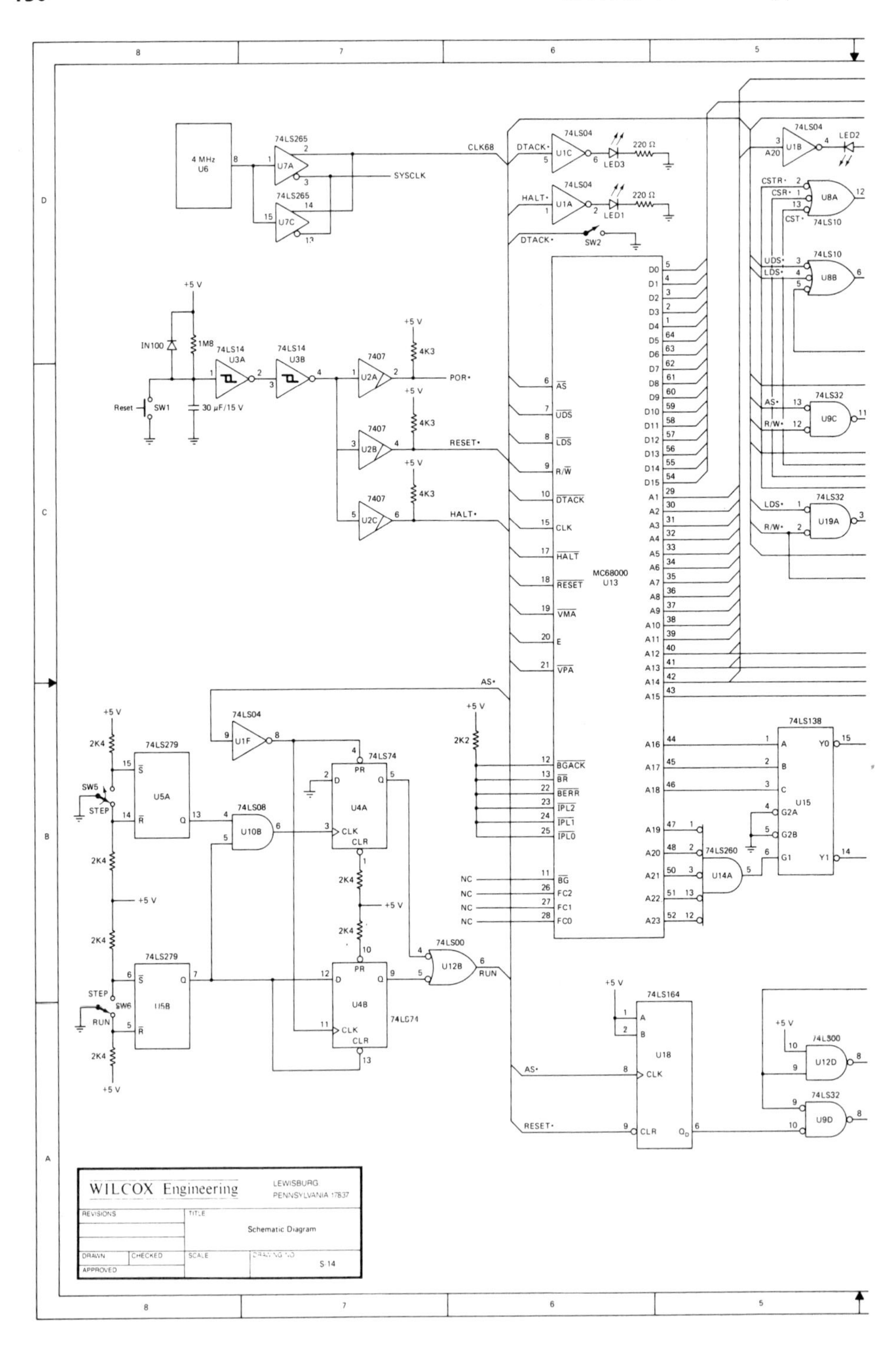
4 MHz
U6
74LS265
U7A
U7C
CLK68
SYSCLK
74LS04
DTACK•
U1C
LED3
220 Ω
HALT•
U1A
LED1
DTACK•
SW2
+5 V
IN100
1M8
74LS14
U3A
U3B
Reset
SW1
30 μF/15 V
7407
U2A
U2B
U2C
4K3
POR•
RESET•
HALT•
MC68000
U13
AS
UDS
LDS
R/W
DTACK
CLK
HALT
RESET
VMA
E
VPA
BGACK
BR
BERR
IPL2
IPL1
IPL0
BG
FC2
FC1
FC0
NC
2K2
AS•
74LS279
U5A
U5B
2K4
SW5
STEP
SW6
RUN
74LS08
U10B
74LS74
U4A
U4B
PR
CLR
74LS00
U12B
RUN
74LS164
U18
RESET•
74LS138
U15
74LS260
U14A
74LS10
U8A
U8B
CSTR•
CSR•
CST•
UDS•
LDS•
74LS32
U9C
U19A
R/W•
A20
U1B
LED2
U12D
U9D
WILCOX Engineering
LEWISBURG
PENNSYLVANIA 17837
REVISIONS
TITLE
Schematic Diagram
DRAWN
CHECKED
SCALE
DRAWING NO
S-14
APPROVED

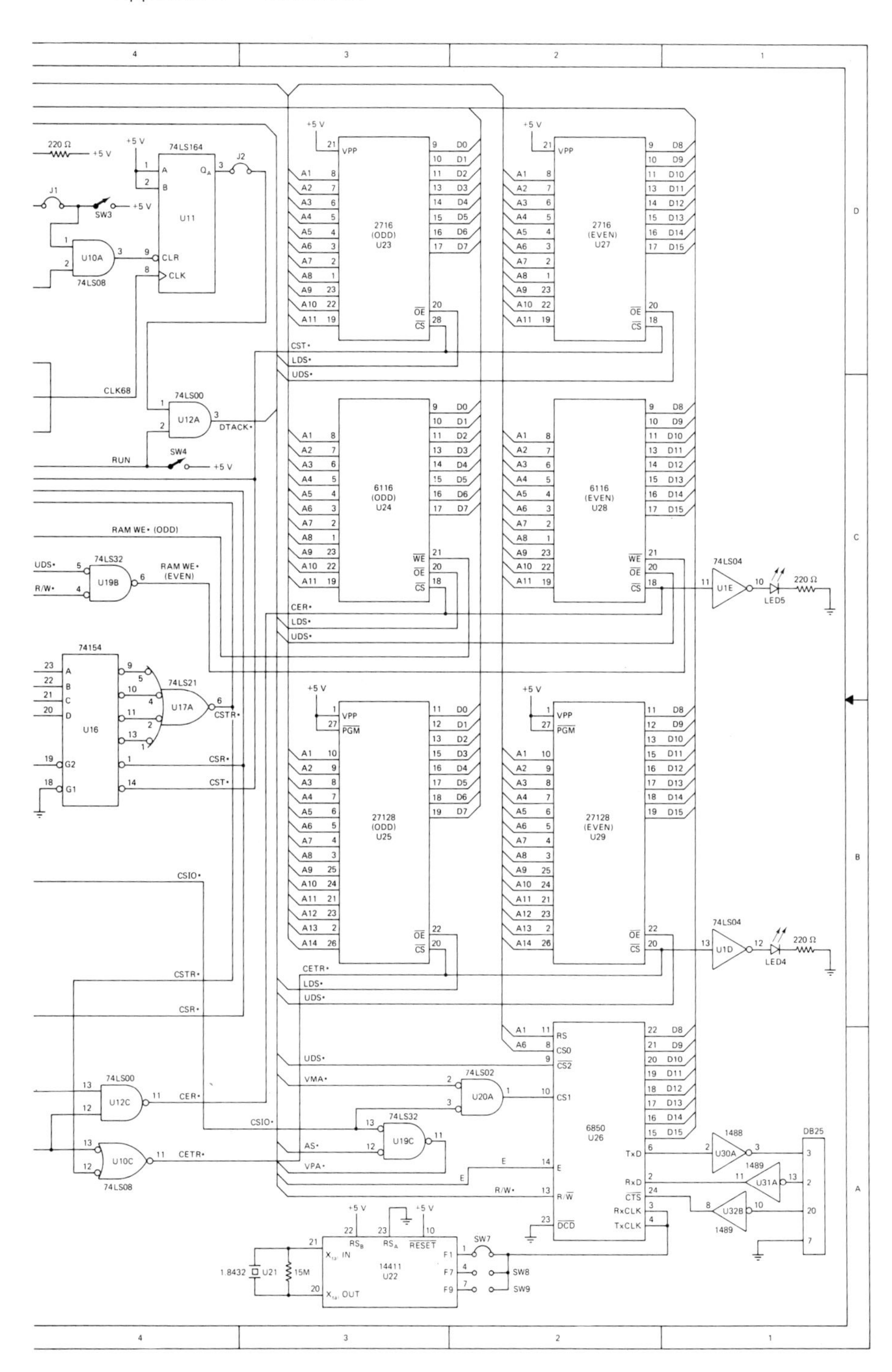
74LS164
U11
74LS08
U10A
74LS00
U12A
DTACK•
RUN
SW4
SW3
J1
J2
CLK68
2716 (ODD) U23
2716 (EVEN) U27
6116 (ODD) U24
6116 (EVEN) U28
27128 (ODD) U25
27128 (EVEN) U29
6850 U26
CST•
LDS•
UDS•
CER•
CETR•
RAM WE• (ODD)
RAM WE• (EVEN)
74LS32
U19B
74154
U16
74LS21
U17A
CSTR•
CSR•
CSIO•
74LS04
U1E
LED5
U1D
LED4
220 Ω
74LS02
U20A
U19C
U12C
U10C
VMA•
AS•
VPA•
R/W•
1488
U30A
1489
U31A
U32B
DB25
14411
U22
1.8432
U21
15M
SW7
SW8
SW9
TxD
RxD
CTS
RxCLK
TxCLK
DCD

APPENDIX D

MM58167A Clock Data Sheet

July 1984

MM58167A Microprocessor Real Time Clock

General Description

The MM58167A is a low threshold metal gate CMOS circuit that functions as a real time clock in bus oriented microprocessor systems. The device includes an addressable real time counter, 56 bits of RAM, and two interrupt outputs. A $\overline{\text{POWER DOWN}}$ input allows the chip to be disabled from the rest of the system for standby low power operation. The time base is a 32,768 Hz crystal oscillator.

Features

- Microprocessor compatible (8-bit data bus)
- Milliseconds through month counters
- 56 bits of RAM with comparator to compare the real time counter to the RAM data
- 2 INTERRUPT OUTPUTS with 8 possible interrupt signals
- $\overline{\text{POWER DOWN}}$ input that disables all inputs and outputs except for one of the interrupts
- Status bit to indicate rollover during a read
- 32,768 Hz crystal oscillator
- Four-year calendar (no leap year)
- 24-hour clock

Functional Description

Real Time Counter

The real time counter is divided into 4-bit digits with 2 digits being accessed during any read or write cycle. Each digit represents a BCD number and is defined in Table I. Any unused bits are held at a logical zero during a read and ignored during a write. An unused bit is any bit not necessary to provide a full BCD number. For example tens of hours cannot legally exceed the number 2, thus only 2 bits are necessary to define the tens of hours. The other 2 bits in the tens of hours digit are unused. The unused bits are designated in Table I as dashes.

The addressable portion of the counter is from milliseconds to months. The counter itself is a ripple counter. The ripple delay is less than 60 μs above 4.0V and 300 μs at 2.0V.

RAM

56 bits of RAM are contained on-chip. These can be used for any necessary power down storage or as an alarm latch for comparison to the real time counter. The data in the RAM can be compared to the real time counter on a digit basis. The only digits that are not compared are the unit ten thousandths of seconds and tens of days of the week (these are unused in the real time counter). If the two most significant bits of any RAM digit are ones, then this RAM location will always compare.

The RAM is formatted the same as the real time counter, 4 bits per digit, 14 digits, however there are no unused bits. The unused bits in the real time counter will compare only to zeros in the RAM.

Interrupts and Comparator

There are two interrupt outputs. The first and most flexible is the INTERRUPT OUTPUT (a true high signal). This output can be programmed to provide 8 different output signals. They are: 10 Hz, 1 Hz, once per minute, once per hour, once a day, once a week, once a month, and when a RAM/real time counter comparison occurs. To enable the output a one is written into the interrupt control register at the bit location corresponding to the desired output frequency *(Figure 1)*. Once one or more bits have been set in the interrupt control register, the corresponding counter's rollover to its reset state will clock the interrupt status register and cause the interrupt output to go high. To reset the interrupt and to identify which frequency caused the interrupt, the interrupt status register is read. Reading this register places the contents of the status register on the data bus. The interrupting frequency will be identified by a one in the respective bit position. Removing the read will reset the interrupt.

The second interrupt is the $\overline{\text{STANDBY INTERRUPT}}$ (open drain output, active low). This interrupt occurs when enabled and when a RAM/real time counter comparison occurs. The $\overline{\text{STANDBY INTERRUPT}}$ is enabled by writing a one on the D0 line at address 16_H or disabled by writing a zero on the D0 line. This interrupt is not triggered by the edge of the compare signal, but rather by the level. Thus if the compare is enabled when the $\overline{\text{STANDBY INTERRUPT}}$ is enabled, the interrupt will turn on immediately.

Connection Diagram

Dual-In-Line Package

Pin	Signal	Pin	Signal
1	$\overline{\text{CS}}$	24	V_{DD}
2	$\overline{\text{RD}}$	23	$\overline{\text{POWER DOWN}}$
3	$\overline{\text{WR}}$	22	D7
4	$\overline{\text{RDY}}$	21	D6
5	A0	20	D5
6	A1	19	D4
7	A2	18	D3
8	A3	17	D2
9	A4	16	D1
10	OSC IN	15	D0
11	OSC OUT	14	$\overline{\text{STANDBY INTERRUPT}}$
12	V_{SS}	13	INTERRUPT OUTPUT

MM58167A

TOP VIEW

TL/F/6148-1

TRI-STATE® is a registered trademark of National Semiconductor Corp.

Absolute Maximum Ratings

Voltage at All Pins	$V_{SS}-0.3V$ to $V_{DD}+0.3V$	V_{DD}-V_{SS}	6.0V
Operating Temperature	0°C to 70°C	Lead Temperature (Soldering, 10 seconds)	300°C
Storage Temperature	−65°C to 150°C		

Electrical Characteristics $V_{SS}=0V, 0°C \leq T_A \leq 70°C$

Parameter	Conditions	Min	Max	Units
Supply Voltage				
V_{DD}	Outputs Enabled	4.0	5.5	V
V_{DD}	$\overline{POWER}$ $\overline{DOWN}$ Mode	2.0	5.5	V
Supply Current				
I_{DD}, Static	Outputs TRI-STATE® $f_{IN}=DC, V_{DD}=5.5V$		10	μA
I_{DD}, Dynamic	Outputs TRI-STATE $f_{IN}=32$ kHz, $V_{DD}=5.5V$ $V_{IH} \geq V_{DD}-0.3V$ $V_{IL} \leq V_{SS}+0.3V$		20	μA
I_{DD}, Dynamic	Outputs TRI-STATE $f_{IN}=32$ kHz, $V_{DD}=5.5V$ $V_{IH}=2.0V, V_{IL}=0.8V$		5	mA
Input Voltage				
Logical Low		0.0	0.8	V
Logical High		2.0	V_{DD}	V
Input Leakage Current	$V_{SS} \leq V_{IN} \leq V_{DD}$	−1	1	μA
Output Impedance	I/O and INTERRUPT OUT			
Logical Low	$V_{DD}=4.5V, I_{OL}=1.6$ mA		0.4	V
Logical High	$V_{DD}=4.5V, I_{OH}=-400$ μA	2.4		V
	$I_{OH}=-10$ μA	0.8 V_{DD}		V
TRI-STATE	$V_{SS} \leq V_{OUT} \leq V_{DD}$	−1	1	μA
Output Impedance	$\overline{RDY}$ and $\overline{STANDBY}$ $\overline{INTERRUPT}$ (Open Drain Devices)			
Logical Low, Sink	$V_{DD}=4.5V, I_{OL}=1.6$ mA		0.4	V
Logical High, Leakage	$V_{OUT} \leq V_{DD}$		10	μA

Functional Description (Continued)

TABLE I. Real Time Counter Format

Counter Addressed		Units D0	D1	D2	D3	Max BCD Code	Tens D4	D5	D6	D7	Max BCD Code
1/10,000 of Seconds	(00_H)	-	-	-	-	0	D4	D5	D6	D7	9
Hundredths and Tenths Sec	(01_H)	D0	D1	D2	D3	9	D4	D5	D6	D7	9
Seconds	(02_H)	D0	D1	D2	D3	9	D4	D5	D6	-	5
Minutes	(03_H)	D0	D1	D2	D3	9	D4	D5	D6	-	5
Hours	(04_H)	D0	D1	D2	D3	9	D4	D5	-	-	2
Day of the Week	(05_H)	D0	D1	D2	-	7	-	-	-	-	0
Day of the Month	(06_H)	D0	D1	D2	D3	9	D4	D5	-	-	3
Month	(07_H)	D0	D1	D2	D3	9	D4	-	-	-	1

(−) indicates unused bits

Functional Description (Continued)

TABLE II. Address Codes and Functions

A4	A3	A2	A1	A0	Function
0	0	0	0	0	Counter—Ten Thousandths of Seconds
0	0	0	0	1	Counter—Hundredths and Tenths of Seconds
0	0	0	1	0	Counter—Seconds
0	0	0	1	1	Counter—Minutes
0	0	1	0	0	Counter—Hours
0	0	1	0	1	Counter—Day of Week
0	0	1	1	0	Counter—Day of Month
0	0	1	1	1	Counter—Month
0	1	0	0	0	RAM—Ten Thousandths of Seconds
0	1	0	0	1	RAM—Hundredths and Tenths of Seconds
0	1	0	1	0	RAM—Seconds
0	1	0	1	1	RAM—Minutes
0	1	1	0	0	RAM—Hours
0	1	1	0	1	RAM—Day of Week
0	1	1	1	0	RAM—Day of Month
0	1	1	1	1	RAM—Months
1	0	0	0	0	Interrupt Status Register
1	0	0	0	1	Interrupt Control Register
1	0	0	1	0	Counters Reset
1	0	0	1	1	RAM Reset
1	0	1	0	0	Status Bit
1	0	1	0	1	GO Command
1	0	1	1	0	$\overline{\text{STANDBY}}$ $\overline{\text{INTERRUPT}}$
1	1	1	1	1	Test Mode

All others unused

Functional Description (Continued)

The comparator is a cascaded exclusive NOR. Its output is latched 61 μs after the rising edge of the 1 kHz clock signal (input to the ten thousandths of seconds counter). This allows the counter to ripple through before looking at the comparator. For operation at less than 4.0V, the thousandths of seconds counter should not be included in a compare because of the possibility of having a ripple delay greater than 61 μs. (For output timing see Interrupt Timing.)

Power Down Mode

The POWER DOWN input is essentially a second chip select. It disables all inputs and outputs except for the STANDBY INTERRUPT. When this input is at a logical zero, the device will not respond to any external signals. It will, however, maintain timekeeping and turn on the STANDBY INTERRUPT if programmed to do so. (The programming must be done before the POWER DOWN input goes to a logical zero.) When switching V_{DD} to the standby or power down mode, the POWER DOWN input should go to a logical zero at least 1 μs before V_{DD} is switched. When switching V_{DD} all other inputs must remain between $V_{SS} - 0.3V$ and $V_{DD} + 0.3V$. When restoring V_{DD} to the normal operating mode, it is necessary to insure that all other inputs are at valid levels before switching the POWER DOWN input back to a logical one. These precautions are necessary to insure that no data is lost or altered when changing to or from the power down mode.

Counter and RAM Resets; GO Command

The counters and RAM can be reset by writing all 1's (FF) at address 12_H or 13_H respectively.

A write pulse at address 15_H will reset the thousandths, hundredths, tenths, units, and tens of seconds counters. This GO command is used for precise starting of the clock. The data on the data bus is ignored during the write. If the seconds counter is at a value greater than 39 when the GO is issued, the minute counter will increment; otherwise the minute counter is unaffected. This command is not necessary to start the clock, but merely a convenient way to start precisely at a given minute.

Status Bit

The status bit is provided to inform the user that the clock is in the process of rolling over when a counter is read. The status bit is set if this 1 kHz clock occurs during or after any counter read. This tells the user that the clock is rippling through the real time counter. Because the clock is rippling, invalid data may be read from the counter. If the status bit is set following a counter read, the counter should be reread.

The status bit appears on D0 when address 14_H is read. All the other data lines will be zero. The bit is set when a logical one appears. This bit should be read every time a counter read or after a series of counter reads are done. The trailing edge of the read at address 14_H will reset the status bit.

Oscillator

The oscillator used is the standard Pierce parallel resonant oscillator. Externally, 2 capacitors, a 20 MΩ resistor and the crystal are required. The 20 MΩ resistor is connected between OSC IN and OSC OUT to bias the internal inverter in the linear region. For micropower crystals a resistor in series with the oscillator output may be necessary to insure the crystal is not overdriven. This resistor should be approximately 200 kΩ. The capacitor values should be typically 20 pF–25 pF. The crystal frequency is 32,768 Hz.

The oscillator input can be externally driven, if desired. In this case the output should be left floating and the input levels should be within 0.3V of the supplies.

A ground line or ground plane between pins 9 and 10 may be necessary to prevent interference of the oscillator by the A4 address.

Control Lines

The READ, WRITE, and CHIP SELECT signals are active low inputs. The READY signal is an open drain output. At the start of each read or write cycle the READY line (open drain) will pull low and will remain low until valid data from a chip read appears on the bus or data on the bus is latched in during a write. READ and WRITE must be accompanied by a CHIP SELECT (see *Figures 3 and 4* for read and write cycle timing).

During a read or write, address bits must not change while chip select and control strobes are low.

Test Mode

The test mode is merely a mode for production testing. It allows the counters to count at a higher than normal rate. In this mode the 32 kHz oscillator input is connected directly to the ten thousandths of seconds counter. The chip select and write lines must be low and the address must be held at $1F_H$.

Functional Description (Continued)

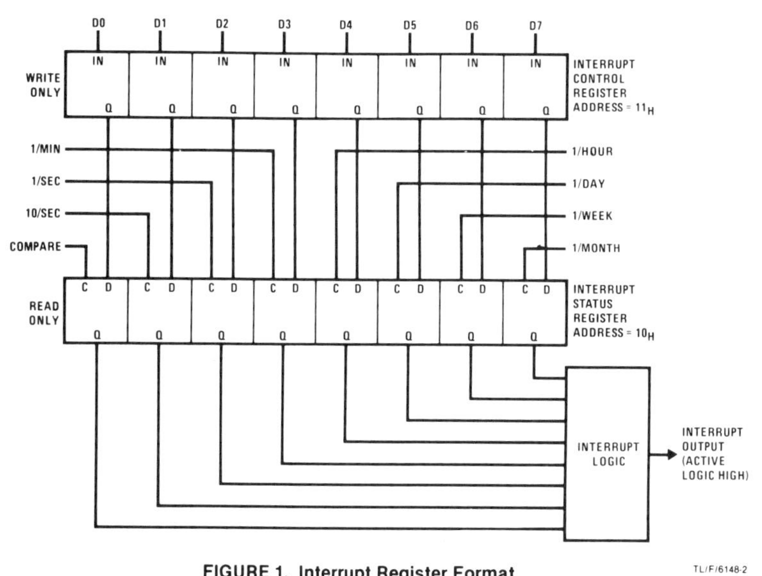

FIGURE 1. Interrupt Register Format

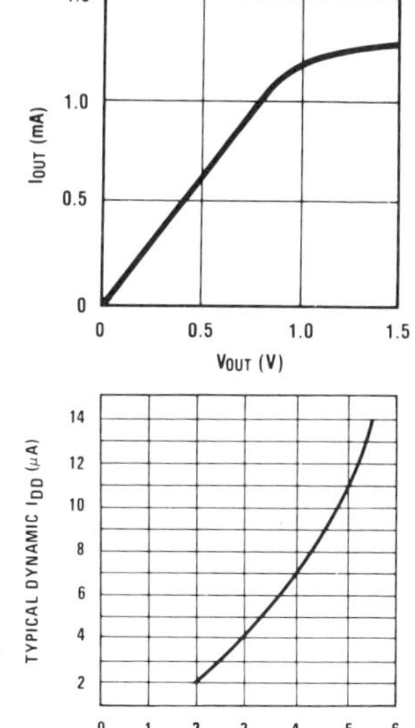

FIGURE 2. Typical Supply Current vs Supply Voltage During Power Down

Interrupt Timing $0°C \leq T_A \leq 70°C$, $4.5V \leq V_{DD} \leq 5.5V$, $V_{SS} = 0V$

Sym	Parameter	Min	Max	Units
t_{INTON}	Status Register Clock to INTERRUPT OUTPUT (Pin 13) High (Note 1)		5	μs
t_{SBYON}	Compare Valid to $\overline{\text{STANDBY}}$ $\overline{\text{INTERRUPT}}$ (Pin 14) Low (Note 1)		5	μs
t_{INTOFF}	Trailing Edge of Status Register Read to INTERRUPT OUTPUT Low		5	μs
t_{SBYOFF}	Trailing Edge of Write Cycle (D0 = 0; Address = 16_H) to $\overline{\text{STANDBY}}$ $\overline{\text{INTERRUPT}}$ Off (High Impedance State)		5	μs

Note 1: The status register clocks are: the corresponding counter's rollover to its reset state or the compare becoming valid. The compare becomes valid 61 μs after the 1/10,000 of a second counter is clocked, if the real time counter data matches the RAM data.

Read Cycle Timing $0°C \leq T_A \leq 70°C$, $4.5V \leq V_{DD} \leq 5.5V$, $V_{SS} = 0V$

Sym	Parameter	Min	Max	Units
t_{AR}	Address Bus Valid to Read Strobe	100		ns
t_{CSR}	Chip Select to Read Strobe	0		ns
t_{RRY}	Read Strobe to Ready Strobe		150	ns
t_{RYD}	Ready Strobe to Data Valid		800	ns
t_{AD}	Address Bus Valid to Data Valid		1050	ns
t_{RH}	Data Hold Time From Trailing Edge of Read Strobe	0		ns
t_{HZ}	Trailing Edge of Read Strobe to TRI-STATE Mode		250	ns
t_{RYH}	Read Hold Time after Ready Strobe	0		ns
t_{RA}	Address Bus Hold Time from Trailing Edge of Read Strobe	50		ns
t_{RYDV}	Rising Edge of Ready to Data Valid		100	ns

Note 2: If $t_{AR} = 0$ and Chip Select, Address Valid or Read are coincident then they must exist for 1050 ns.

Write Cycle Timing $0°C \leq T_A \leq 70°C$, $4.5V \leq V_{DD} \leq 5.5V$, $V_{SS} = 0V$

Sym	Parameter	Min	Max	Units
t_{AW}	Address Valid to Write Strobe	100		ns
t_{CSW}	Chip Select to Write Strobe	0		ns
t_{DW}	Data Valid before Write Strobe	100		ns
t_{WRY}	Write Strobe to Ready Strobe		150	ns
t_{RY}	Ready Strobe Width		800	ns
t_{RYH}	Write Hold Time after Ready Strobe	0		ns
t_{WD}	Data Hold Time after Write Strobe	110		ns
t_{WA}	Address Hold Time after Write Strobe	50		ns

Note 3: If data changes while $\overline{CS}$ and $\overline{WR}$ are low, then they must remain coincident for 1050 ns after the data change to ensure a valid write.

Data bus loading is 100 pF.

Ready output loading is 50 pF and 3 kΩ pull-up.

Input and output AC timing levels:
Logical one = 2.0V
Logical zero = 0.8V

Read and Write Cycle Timing Diagrams

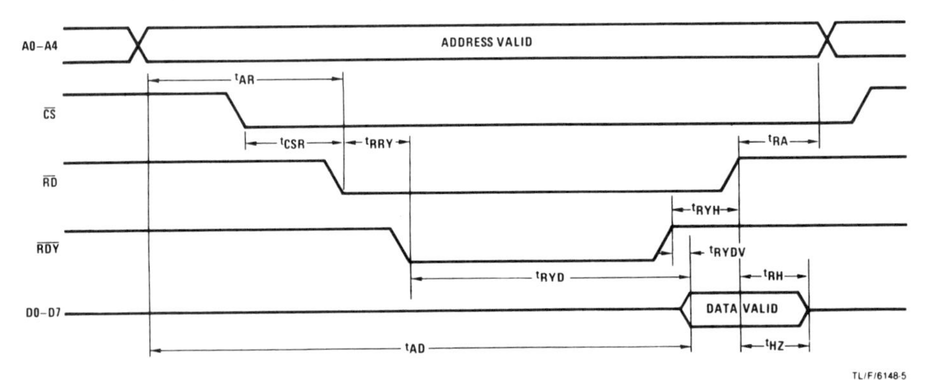

FIGURE 3. Read Cycle Timing

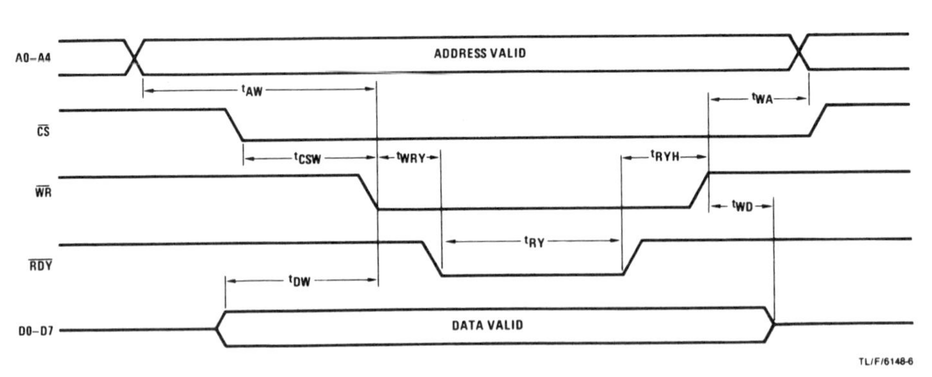

FIGURE 4. Write Cycle Timing

Typical Applications

R1 = 20 MΩ ± 20%
C1 = 6 pF – 36 pF
R2 to be selected based on crystal used.

TL/F/6148-7

Note 4: A ground line or ground plane guard trace should be included between pins 9 and 10 to insure the oscillator is not disturbed by the address line.

FIGURE 5. Typical Connection Diagram

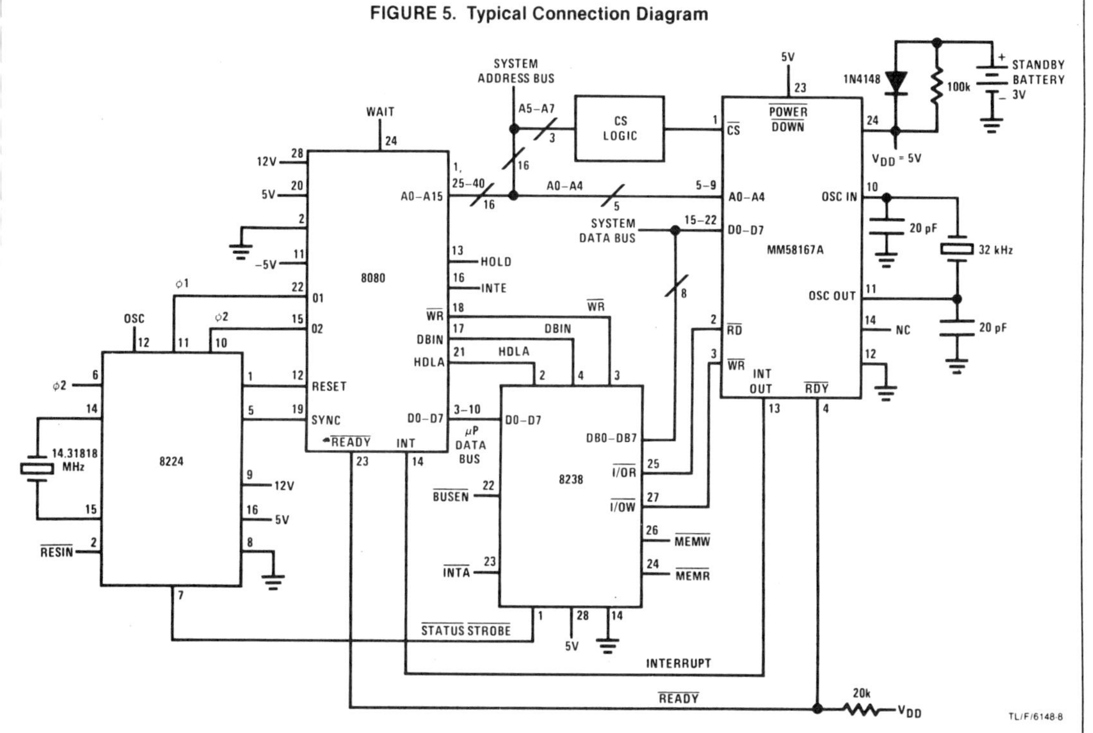

Note 5: Must use 8238 or equivalent logic to insure advanced I/OW pulse; so that the ready output of the MM58167A is valid by the end of **φ2** during the T2 microcycle.

Note 6: $t_{\phi 2} \geq t_{RS8080} + t_{DL8238} + t_{WRY58167A}$.

FIGURE 6. 8080 System Interface with Battery Backup

Block Diagram

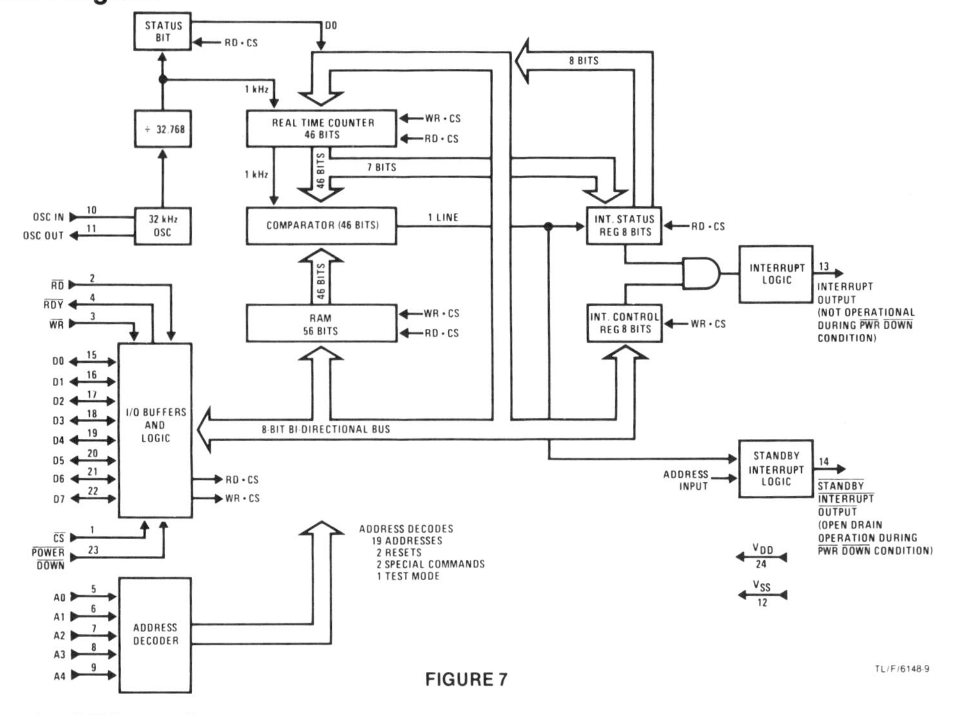

FIGURE 7

Physical Dimensions inches (millimeters)

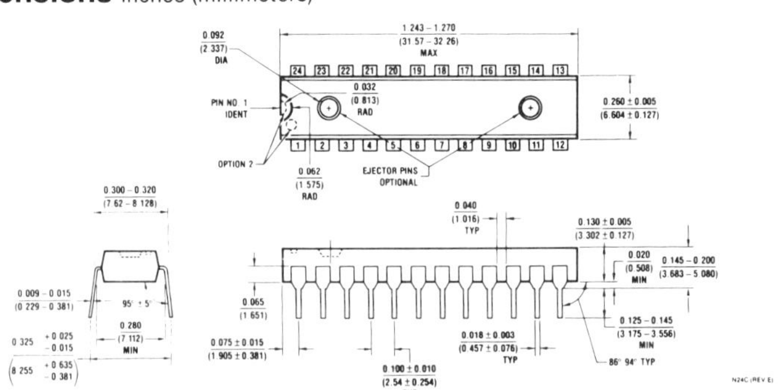

Molded Dual-In-Line Package (N)
Order Number MM58167AN
NS Package Number N24C

LIFE SUPPORT POLICY

NATIONAL'S PRODUCTS ARE NOT AUTHORIZED FOR USE AS CRITICAL COMPONENTS IN LIFE SUPPORT DEVICES OR SYSTEMS WITHOUT THE EXPRESS WRITTEN APPROVAL OF THE PRESIDENT OF NATIONAL SEMICONDUCTOR CORPORATION. As used herein:

1. Life support devices or systems are devices or systems which, (a) are intended for surgical implant into the body, or (b) support or sustain life, and whose failure to perform, when properly used in accordance with instructions for use provided in the labeling, can be reasonably expected to result in a significant injury to the user.
2. A critical component is any component of a life support device or system whose failure to perform can be reasonably expected to cause the failure of the life support device or system, or to affect its safety or effectiveness.

National Semiconductor Corporation
2900 Semiconductor Drive
Santa Clara, California 95051
Tel: (408) 721-5000
TWX: (910) 339-9240

National Semiconductor GmbH
Furstenriederstrasse Nr. 5
D8000 Munchen 21
West Germany
Tel: (089) 5 60 12-0
Telex: 522772

NS Japan K.K.
4-403 Ikebukuro, Toshima-ku
Tokyo 171, Japan
Tel: (03)988-2131
FAX: 011-81-3-988-1700

National Semiconductor Hong Kong Ltd.
Southeast Asia Marketing
N.H.K. Sales
Austin Tower, 4th Floor
22-26 Austin Avenue
Tsimshatsui, Kowloon, Hong Kong
Telephone: 3-7231290, 3-7243645
Cable: NSSEAMKTG
Telex: 52996 NSSEA HX

National Semiconductores Do Brasil Ltda.
Avda Brigadeiro Faria Lima 830
8 ANDAR
01452 Sao Paulo, Brasil
Tel: 212-1181
Telex: 1131931 NSBR

NS Electronics Pty. Ltd.
Cnr. Stud Rd. & Mtn. Highway
Bayswater, Victoria 3153
Australia
Tel: 03-729-6333
Telex: 32096

Index